D0347764

ANIMAL FARM

GEORGE ORWELL

WORKBOOK BY DAVID GRANT

YORK PRESS
322 Old Brompton Road, London SW5 9JH

PEARSON EDUCATION LIMITED
Edinburgh Gate, Harlow,
Essex CM20 2JE, United Kingdom
Associated companies, branches and representatives throughout the world

First published 2015

10 9 8 7 6 5 4 3 2 1

ISBN 978-1-2921-0078-4

Illustrations by John Rabou; and Moreno Chiacchiera (page 54 only)

Phototypeset by Carnegie Book Production
Printed in Slovakia

Photo credits: Andrew Rowland/Shutterstock for page 14 bottom / natalia bulatova/Shutterstock for page 17 top / Regien Paassen/Shutterstock for page 19 top / Lintra/Shutterstock for page 22 bottom / pbombaert/Shutterstock for page 27 top / © iStock/Akabei for page 31 top / © Peter Coleman for page 49 top / Hulton Archive/Getty Images for page 50 bottom / English School/Getty Images for page 52 (Marx) / Hulton Archive/Getty Images for page 52 (Lenin) / © Corbis/Bettman for page 52 (Trotsky)/ Alamy/Everett Collection Historical for page 52 (Stalin) / vagant/ Shutterstock for page 53 top

CONTENTS

PART ONE: GETTING STARTED

PART TWO: PLOT AND ACTION

PART THREE: CHARACTERS

PART FOUR: THEMES, CONTEXTS AND SETTINGS

PART FIVE: FORM, STRUCTURE AND LANGUAGE

PART SIX: PROGRESS BOOSTER

PART ONE: GETTING STARTED

Preparing for assessment

HOW WILL I BE ASSESSED ON MY WORK ON *ANIMAL FARM*?

All exam boards are different, but whichever course you are following, your work will be examined through these four Assessment Objectives:

Assessment Objectives	Wording	Worth thinking about ...
A01	Read, understand and respond to texts. Students should be able to: • maintain a critical style and develop an informed personal response • use textual references, including quotations, to support and illustrate interpretations.	• How well do I know what happens, what people say, do, etc? • What do *I* think about the key ideas in the text? • How can I support my viewpoint in a convincing way? • What are the best quotations to use and when should I use them?
A02	Analyse the language, form and structure used by a writer to create meanings and effects, using relevant subject terminology where appropriate.	• What specific things does the writer 'do'? What choices has Orwell made? (Why this particular word, phrase or paragraph here? Why does this event happen at this point?) • What effects do these choices create? Suspense? Characterisation? Symbolic significance?
A03	Show understanding of the relationships between texts and the contexts in which they were written.	• What can I learn about society from the book? (What does it tell me about people's political concerns in Orwell's day, for example?) • What was society like in Orwell's time? Can I see it reflected in the story?
A04	Use a range of vocabulary and sentence structures for clarity, purpose and effect, with accurate spelling and punctuation.	• How accurately and clearly do I write? • Are there small errors of grammar, spelling and punctuation I can get rid of?

Look out for the Assessment Objective labels throughout your York Notes Workbook – these will help to focus your study and revision!

The text used in this Workbook is the Heinemann New Windmill edition, 1994.

How to use your York Notes Workbook

There are lots of ways your Workbook can support your study and revision of *Animal Farm*. There is no 'right' way – choose the one that suits your learning style best.

1) Alongside the York Notes Study Guide and the text	2) As a 'stand-alone' revision programme	3) As a form of mock-exam
Do you have the York Notes Study Guide for *Animal Farm*? The contents of your Workbook are designed to match the sections in the Study Guide, so with the text to hand you could: • read the relevant section(s) of the Study Guide and any part of the text referred to; • complete the tasks in the same section in your Workbook.	Think you know *Animal Farm* well? Why not work through the Workbook systematically, either as you finish chapters, or as you study or revise certain aspects in class or at home. You could make a revision diary and allocate particular sections of the Workbook to a day or week.	Prefer to do all your revision in one go? You could put aside a day or two and work through the Workbook, page by page. Once you have finished, check all your answers in one go! This will be quite a challenge, but it may be the approach you prefer.

HOW WILL THE WORKBOOK HELP YOU TEST AND CHECK YOUR KNOWLEDGE AND SKILLS?

Parts Two to **Five** offer a range of tasks and activities:

These fun and quick-to-complete tasks check your basic knowledge of the text

PART TWO: PLOT AND ACTION

Chapter 3

QUICK TEST

1. **Complete** these sentences describing how the animals share the work of the harvest.
   a) *The pigs ...*
   b) *The horses ...*
   c) *The hens and ducks ...*
   d) *All the other animals ...*
2. What is Boxer's answer to every problem?
   a) 'Four legs good, two legs bad' ☐
   b) 'I will work harder!' ☐
   c) 'We should ask Napoleon and Snowball.' ☐
   d) To practise writing the letters A to D ☐

THINKING MORE DEEPLY

3. Write **one** or **two sentences** in response to each of these questions:
   a) Benjamin will not say whether he thinks things have improved now Jones has gone. All he will say is 'Donkeys live a long time' (p. 18). What is he implying?
   b) The pigs now openly take all the milk and apples for themselves. What does this suggest about the pigs and the way the farm is organised?
   c) Squealer explains that Jones will come back if the pigs do not have all the milk and apples. How effective is the argument in persuading the other animals?

These more open questions challenge you to show your understanding

16 Animal Farm

PART TWO: PLOT AND ACTION

EXAM PREPARATION: WRITING ABOUT TENSION A02

Reread the section of Chapter 3 from *'None of the other animals on the farm could get further than the letter A'* (p. 20) to *'surely there is none among you who wants to see Jones come back?'* (p. 22).

Question: How does Orwell create tension in this passage?

Think about:

- How the animals behave and what they say
- How Orwell suggests that the pigs may not be acting in the farm's best interests

4. Complete this table:

Point/detail	Evidence	Effect or explanation
1: *The pigs' intelligence allows them to manipulate and exploit the farm animals just as Jones did.*	*'The birds did not understand Snowball's long words, but they accepted his explanation'*	*The pigs are effectively able to control the other animals' thoughts, attitudes and decisions.*
2: *The pigs take all the apples and milk for themselves.*		
3: *Squealer's dishonest explanation suggests that the pigs will continue to lie and to cheat to get their own way.*		

5. Write up **point 1** into a **paragraph** below in your own words. Remember to include what you infer from the evidence, or the writer's effects.
6. Now, choose **one** of your **other points** and write it out as another **paragraph** here:

PROGRESS LOG [tick the correct box] Needs more work ☐ Getting there ☐ Under control ☐

Animal Farm

This task focuses on a key character, theme, technique, idea or relationship and helps you plan and write up paragraphs from an essay

A clear, quick way to record your progress visually

Each Part ends with a **Practice task** to extend your revision:

An exam-style task is provided at the end of each section for you to practise a full essay

PART TWO: PLOT AND ACTION

Practice task

1. First, **read** this **exam-style** task:

Read the section from *'November came, with raging south-west winds'* (Ch. 6, p. 45) to the end of Chapter 6.

Question: In this passage, what methods does Orwell use to present characters and events? Refer closely to the passage in your answer.

2. Begin by circling the **key words** in the **question** above.

3. Now complete this table, noting down **three or four key points** with **evidence** and the **effect created**.

Point	Evidence/quotation	Meaning or effect

4. **Draft your response.** Use the space below for your first paragraph(s) and then continue onto a sheet of paper.

Start: *In this extract, Orwell describes the animals' reaction to the destruction of the windmill by a storm ...*

PROGRESS LOG [tick the correct box] Needs more work ☐ Getting there ☐ Under control ☐

32 Animal Farm

A plain table is provided for you to fill in with your own ideas

The first sentence of the essay is provided for you to use as a prompt to start a full-length essay

Part Six: Progress Booster helps you test your own key writing skills:

PART SIX: PROGRESS BOOSTER

Sample answers (A01) (A02) (A03) (A04)

OPENING PARAGRAPHS

Here is the task from the previous page:

Question: *How are the pigs depicted in this extract and how do the animals respond to them?*

A sample of a student's writing challenges you to judge its strengths and weaknesses

Mid level

Lower level

1. Which comment belongs to which answer? Match the paragraph (A or B) to the expert's feedback (1 or 2).

Student A: ________ Student B: ________

2. Now it's your turn. Write the opening paragraph to this task on a separate sheet of paper:

Read from *'It was about this time'* (Ch. 6, p. 42) to *'no complaint was made about that either.'* (Ch. 10, p. 43).

Question: *How are the pigs depicted in this extract and how do the animals respond to them?*

Remember:

- Introduce the topic in general terms, perhaps **explaining** or **'unpicking'** the key **words** or **ideas** in the task (such as 'depicted').
- Mention the **different possibilities** or ideas that you are going to address.
- Use the **author's name.**

74 Animal Farm

PART SIX: PROGRESS BOOSTER

WRITING ABOUT TECHNIQUES

Here are two paragraphs in response to a different task, where the students have focused on the writer's techniques. The task is:

Read from *'It was about this time'* (Ch. 6, p. 42) to *'no complaint was made about that either.'* (Ch. 10, p. 43).

Question: *What techniques does Orwell use to show how the pigs control the other animals?*

Student A

Orwell gives Squealer a sequence of rhetorical questions which the animals do not, or perhaps dare not, interrupt. This is because the questions need no answer: the answers are obvious. None of the animals 'wishes to see Jones back' and so Squealer exploits their fears to justify the pigs' actions and uses the rhetorical questions to make the pigs' decisions seem inevitable. Their authority cannot be challenged.

Student B

Squealer uses a series of rhetorical questions to persuade the animals that the pigs need to sleep in beds so that they can manage 'all the brainwork' that they do. These rhetorical questions end with 'Surely none of you wishes to see Jones back?' which shows how Squealer uses fear to frighten the animals into silence. This is what Squealer does very well in the novel.

Expert viewpoint 1: This higher-level response explores the immediate impact of the language techniques used by Squealer and explores how they are effective and their wider implications in the context of the novel. The penultimate sentence is a little long, but links ideas very successfully and is effectively summarised in the final short sentence. Higher level

Expert viewpoint 2. This mid-level response highlights Squealer's use of rhetorical questions on the other animals and comments on the effect of one. However, this comment could be further developed and focus more broadly on why Orwell chose to use this technique at this point in the novel. Mid level

3. Which comment belongs to which answer? Match the paragraph (A or B) to the examiner feedback (1 or 2).

Student A: ________ Student B: ________

4. Now, take another **aspect** of the passage and write your own paragraph. You could **comment** on one of these aspects:

An expert teacher or marker's view of the student's work will help you understand key skills

An opportunity for you to apply what you have learned to a new point

Don't forget – these are just some examples of the Workbook contents. Inside there is much more to help you revise. For example:

- lots of examples of students' own work at different levels
- help with spelling, punctuation and grammar
- advice and tasks on writing about context
- a full answer key so you can check your answers
- a full-length practice exam task with guidance on what to focus on.

PART TWO: PLOT AND ACTION

Chapter 1

QUICK TEST

1. Why doesn't Mr Jones shut the pop-holes on the hen-houses?

   a) He is too tired. ☐

   b) He is too lazy. ☐

   c) He is too drunk. ☑

2. In this chapter, Major calls the animals together for a meeting. Why? Which of these statements are **TRUE** and which are **FALSE**? Write **'T'** or **'F'** in the boxes:

   a) He wants them to have a meeting every week to share their news and views. [F]

   b) He wants to tell them what he has learned about the nature of life before he dies. [T]

   c) He wants to tell them about a dream he has had. [T]

   d) He wants to tell them that Man is their enemy. [T]

   e) He wants all the animals to leave Manor Farm and move to another, better farm. [F]

   f) He wants them to work towards a rebellion to overthrow the rule of Man. [T]

3. Major gives the animals a set of **rules or principles** by which he thinks they should live. Which of the following accurately sum up his principles? **Tick** all that do and **cross** any that do not:

   a) Whatever goes on two legs is an enemy. ☑

   b) Whatever goes upon four legs and has wings is a friend – except rats. ☒

   c) Whatever goes upon four legs and has wings is a friend. ☑

   d) Animals should not behave like men. ☑

   e) No animal should kill another animal. ☑

   f) No animal should speak to or negotiate with men. ☒

   g) All animals are equal. ☑

4 **Circle** any animals who are **absent** from the meeting.

Bluebell, Jessie and Pitcher *the pigs* *the hens* *the sheep*
[*Moses*] *Clover and Boxer* *Mollie* *the cat* *the rats*
Muriel *Benjamin* *the cows*

THINKING MORE DEEPLY ?

5 Write **one** or **two sentences** in response to each of these questions:

a) How does Major suggest life would change if the animals could 'get rid of Man' (p. 5)?

Major suggests that life would change for the animals if they got 'rid of man' by ranting. "Man does not produce..." This suggests that the animals would be better off working for themselves and that they would never be overworked or hungry.

b) Major declares that 'All men are enemies. All animals are comrades' (p. 5).
How does Orwell immediately suggest that Major's vision might be too optimistic?

When Old Major says this, 'four large rats' appear and seem to catch the eye of the cats & dogs who try to eat them. This suggests that Majors speech is too optimistic & he acts too fast in thinking about the future because already, Orwell shows the irony that the animals will & do find it hard to unify eachother.

c) Orwell tells us that Major is 'highly regarded' (p. 1) by the other animals.
How does Chapter 1 suggest this?

All the animals attened his meeting, consider his dream to their benefits and sing 'Beasts of England' with him.

d) How does Orwell suggest in Chapter 1 that the animals will put Major's ideas into action?

6 Four other animals are introduced in Chapter 1. What do you learn about them? (Write **one** or **two sentences** for each animal below.)

a) Boxer

b) Benjamin

c) Clover

d) Mollie

7 Why do the animals sing 'Beasts of England' at the end of the chapter?

EXAM PREPARATION: WRITING ABOUT MANKIND

A01

Reread the section of Chapter 1 from page 3 *'Now, comrades, what is the nature of this life of ours?'* to page 5 *'Jones ties a brick round their necks and drowns them in the nearest pond.'*

Question: What impression does Orwell create of mankind in Chapter 1?

Think about:

- How mankind treats the animals
- How Major expresses his ideas

8 Complete this table:

Point/detail	Evidence	Effect or explanation
1: *Major describes the lives of farm animals.*	*'Let us face it: our lives are miserable, laborious, and short.'*	*Major's blunt, negative language in this short sentence emphasises how badly farm animals are treated.*
2: *Major explains how mankind exploits animals.*		
3: *Major describes the deaths of farm animals.*		

9 Write up **point 1** into a **paragraph** below in your own words. Remember to include what you infer from the evidence, or the writer's effects.

....

....

....

....

....

10 Now, choose **one** of your **other points** and write it out as another **paragraph** here:

....

....

....

....

....

PROGRESS LOG [tick the correct box] Needs more work ☐ Getting there ☐ Under control ☐

Chapter 2

QUICK TEST

1. Why do the pigs lead the preparations for the rebellion?

   a) They are the bossiest of the animals. ☐
   b) They are the most respected of the animals. ☐
   c) Major is a pig. ☐
   d) They are recognised as the most intelligent of the animals. ☐

2. What is Animalism?

   a) A system of thought based on the teachings of Major ☐
   b) The belief that animals are superior to men ☐
   c) Behaving or speaking in a way which shows prejudiced attitudes towards animals ☐
   d) A religion in which believers worship Major ☐

3. Why does Mr Jones neglect his farm and his animals? Tick **any** of the following reasons which are given in the text:

   a) He is not a very good farmer. ☐
   b) He had lost money and become disheartened. ☐
   c) He drinks heavily. ☐
   d) He is evil. ☐

4. Look at this list of events building up to the animals taking over the farm. They are in the wrong order. Number the events in the order they take place in Chapter 2.

   A Jones and his men whip the animals. ☐
   B Mrs Jones packs a few things and flees the farm. ☐
   C Mr Jones goes to the pub and stays out all night. ☐
   D Jones and his men flee the farm. ☐
   E Everything that Jones used to control or mistreat the animals is thrown down the well. ☐
   F The farm workers forget to feed the animals. ☐
   G The animals retaliate. ☐
   H Mr Jones returns to the farm and goes to sleep. ☐
   I The animals break into the store-shed and help themselves to feed. ☐

THINKING MORE DEEPLY ?

5. In Chapter 2 the reader is introduced to the pigs Napoleon, Snowball and Squealer. What impressions does Orwell give the reader of them?

   a) Napoleon

   b) Snowball

   c) Squealer

6. *Animal Farm* can be read as a political allegory reflecting the events of the Russian Revolution. For example, Major can be compared to Karl Marx whose ideas were the foundation on which the revolution was built. What might Sugarcandy Mountain represent in an **allegorical** reading of the text?

7. Orwell gives us several reasons why Mr Jones neglects his farm and his animals. Is Orwell encouraging the reader to blame Jones or to sympathise with him? Give **one** or **two reasons** for your answer.

8 Look again at the animals' objections to Animalism and the rebellion:

- Some animals call Mr Jones 'Master'.
- Some think it is their duty to be loyal to Mr Jones.
- Some think they will starve if Mr Jones does not feed them.
- Some think the rebellion will not happen in their lifetime.
- Some think the rebellion will happen even if they do nothing.

What do these objections suggest about the animals on Animal Farm?

9 Clover and Boxer are described as the pigs' 'most faithful disciples' (p. 10). What does the word 'disciples' suggest about Clover and Boxer's attitude to the pigs and to Animalism?

10 After the rebellion, the animals all agree that no animal must ever live in the farmhouse. Why do they make this decision?

11 Which, if any, of the commandments have been broken by the end of Chapter 2?

EXAM PREPARATION: WRITING ABOUT THE PIGS

A01

Reread the section of Chapter 2 from page 14 '*"Comrades," said Snowball, "it is half past six and we have a long day before us"*' to the end of the chapter.

Question: What do the details in this passage suggest about the relationship between the pigs and the other animals – and about what might happen in the future?

Think about:

- What the pigs do and say
- What the other animals do and say

⑫ Complete this table:

Point/detail	Evidence	Effect or explanation
1: *The pigs use their intelligence to control and direct the other animals.*	*'The pigs now revealed that during the past three months they had taught themselves to read and write'*	*The word 'revealed' indicates that the pigs have acquired this human skill secretly and their actions may not be honest or fair.*
2: *Orwell clearly implies why the milk has disappeared at the end of the chapter.*		
3: *Orwell suggests that the pigs may not look after the animals as well as Jones did.*		

⑬ Write up **point 1** into a **paragraph** below in your own words. Remember to include what you infer from the evidence, or the writer's effects.

....

....

....

....

....

⑭ Now, choose **one** of your **other points** and write it out as another **paragraph** here:

....

....

....

....

....

PROGRESS LOG [tick the correct box] Needs more work ☐ Getting there ☐ Under control ☐

Chapter 3

QUICK TEST

❶ **Complete** these sentences describing how the animals share the work of the harvest.

a) *The pigs …* ....

b) *The horses …* ....

c) *The hens and ducks …* ....

d) *All the other animals …* ....

❷ What is Boxer's answer to every problem?

a) 'Four legs good, two legs bad' ☐

b) 'I will work harder!' ☐

c) 'We should ask Napoleon and Snowball.' ☐

d) To practise writing the letters A to D ☐

THINKING MORE DEEPLY

❸ Write **one** or **two sentences** in response to each of these questions:

a) Benjamin will not say whether he thinks things have improved now Jones has gone. All he will say is 'Donkeys live a long time' (p. 18). What is he implying?

....

....

....

....

b) The pigs now openly take all the milk and apples for themselves. What does this suggest about the pigs and the way the farm is organised?

....

....

....

....

c) Squealer explains that Jones will come back if the pigs do not have all the milk and apples. How effective is the argument in persuading the other animals?

....

....

....

....

....

EXAM PREPARATION: WRITING ABOUT TENSION

AO2

Reread the section of Chapter 3 from *'None of the other animals on the farm could get further than the letter A'* (p. 20) to *'surely there is none among you who wants to see Jones come back?'* (p. 22).

Question: How does Orwell create tension in this passage?

Think about:

- How the animals behave and what they say
- How Orwell suggests that the pigs may not be acting in the farm's best interests

❹ Complete this table:

Point/detail	Evidence	Effect or explanation
1: *The pigs' intelligence allows them to manipulate and exploit the farm animals just as Jones did.*	*'The birds did not understand Snowball's long words, but they accepted his explanation'*	*The pigs are effectively able to control the other animals' thoughts, attitudes and decisions.*
2: *The pigs take all the apples and milk for themselves.*		
3: *Squealer's dishonest explanation suggests that the pigs will continue to lie and to cheat to get their own way.*		

❺ Write up **point 1** into a **paragraph** below in your own words. Remember to include what you infer from the evidence, or the writer's effects.

....

....

....

....

....

❻ Now, choose **one** of your **other points** and write it out as another **paragraph** here:

....

....

....

....

....

....

PROGRESS LOG [tick the correct box] Needs more work ☐ Getting there ☐ Under control ☐

Chapter 4

QUICK TEST

❶ How does Orwell show the growing popularity of Animalism? **Tick** all the correct answers.

a) A number of other farms have been taken over by animals. ☐

b) The farm animals are becoming more rebellious. ☐

c) Lots of animals are heard singing 'Beasts of England'. ☐

❷ Who is awarded a medal for their actions in the Battle of the Cowshed?

a) Napoleon, Snowball and Boxer ☐

b) Snowball, a sheep and Boxer ☐

c) Napoleon, Boxer and a sheep ☐

❸ Which animals do not take part in the Battle of the Cowshed?

a) Mollie and Napoleon ☐

b) Benjamin and Mollie ☐

c) Snowball and Napoleon ☐

THINKING MORE DEEPLY

❹ Write **one** or **two sentences** in response to each of these questions:

a) For what positive or negative reasons might Napoleon and Snowball be trying to spread Animalism to other farms?

....

....

....

b) Why might Orwell have decided that no humans should die in the Battle of the Cowshed?

....

....

....

c) Do you think Orwell wants the reader to support, or be disturbed by, the Battle of the Cowshed?

....

....

....

....

EXAM PREPARATION: WRITING ABOUT THE BATTLE OF THE COWSHED A01

Reread the section of Chapter 4 from *'All the men were gone except one.'* (p. 26) to *'ready to die for Animal Farm if need be.'* (p. 27).

Question: How has the Battle of the Cowshed affected your response to the animals of Animal Farm?

Think about:

- What the animals do in the battle
- How you respond to the aftermath of the battle

❺ Complete this table:

Point/detail	Evidence	Effect or explanation
1: *Snowball is badly hurt.*	*'from whose wounds the blood was still dripping'*	*The fact that Snowball is injured suggests his dedication to Animalism and the other animals.*
2: *The animals work together very successfully.*		
3: *The animals celebrate their victory.*		

❻ Write up **point 1** into a **paragraph** below in your own words. Remember to include what you infer from the evidence, or the writer's effects.

....

....

....

....

....

❼ Now, choose **one** of your **other points** and write it out as another **paragraph** here:

....

....

....

....

....

PROGRESS LOG [tick the correct box] Needs more work ☐ Getting there ☐ Under control ☐

Chapter 5

QUICK TEST

1. What is the first sign that Napoleon does not approve of Snowball's plans for a windmill?

   a) He urinates on the plans. ☐

   b) He argues eloquently against the idea of a windmill. ☐

   c) He proposes a much better idea. ☐

2. Which pig makes which arguments? **Draw lines** to link each pig with his points of view.

	The animals should concentrate on food production, not windmills.
Napoleon	
	The windmill would save so much labour that the animals could work for just three days a week.
	The animals should get guns to protect themselves from another attack by humans.
Snowball	
	The animals should stir up rebellion on other farms to make all animals more powerful.

THINKING MORE DEEPLY

3. Write **one** or **two sentences** in response to each of these questions:

   a) Why do you think the sheep bleat 'Four legs goods, two legs bad' (p. 29) during Snowball's speech?

   ....

   ....

   ....

   b) Why do you think Napoleon says so little in the debate about whether to build a windmill?

   ....

   ....

   ....

   c) Why do you think Napoleon changes his mind and decides to build the windmill?

   ....

   ....

   ....

EXAM PREPARATION: WRITING ABOUT THE DOGS

Reread the section of Chapter 5 from *'At this there was a terrible baying sound'* (p. 33) to *'there would be no more debates'* (p. 34).

Question: How does Orwell present Napoleon's dogs and the impact they have on the other animals?

Think about:

- The description of the dogs and the reaction of the animals
- The changes Napoleon announces after Snowball has fled

❹ Complete this table:

Point/detail	Evidence	Effect or explanation
1: *Orwell presents the dogs as vicious and dangerous.*	*'a terrible baying sound … enormous dogs … snapping jaws'*	*Focusing on their size and their jaws, Orwell highlights how intimidating the dogs are, and by implication how intimidating Napoleon is too.*
2: *The animals are deeply affected by this incident.*		
3: *Napoleon uses this incident to increase his power.*		

❺ Write up **point 1** into a **paragraph** below in your own words. Remember to include what you infer from the evidence, or the writer's effects.

❻ Now, choose **one** of your **other points** and write it out as another **paragraph** here:

PROGRESS LOG [tick the correct box] Needs more work ☐ Getting there ☐ Under control ☐

Chapter 6

QUICK TEST ✓

1. How does Napoleon encourage the animals to work voluntarily on Sunday afternoons? **Tick** the box with the correct answer.

   a) They will have double rations if they do. ☐

   b) He says the windmill will be finished sooner if they work harder. ☐

   c) They will have half-rations if they do not. ☐

2. Tick **all** the principles and Commandments of Animalism that are broken in this chapter.

   a) Never to use money or have any dealings with humans ☐

   b) Never to drink alcohol ☐

   c) Never to live in the farmhouse ☐

   d) Never to sleep in a bed ☐

   e) Never to kill another animal ☐

THINKING MORE DEEPLY ?

3. Write **one** or **two sentences** in response to each of these questions:

   a) Why do you think Orwell again emphasises how hard Boxer works in this chapter?

   ....

   ....

   ....

   ....

   ....

   b) What is Squealer's main role on the farm?

   ....

   ....

   ....

   ....

   c) Why does Napoleon blame Snowball for the destruction of the windmill?

   ....

   ....

   ....

   ....

EXAM PREPARATION: WRITING ABOUT THE PIGS' CONTROL

A01

Reread the section of Chapter 6 from *'It was about this time that the pigs suddenly moved into the farmhouse'* (p. 42) to *'no complaint was made about that either'* (p. 43).

Question: What does this section reveal about the ways in which the pigs control the other animals?

Think about:

- Clover questioning the pigs' actions
- How the pigs control the other animals

❹ Complete this table:

Point/detail	Evidence	Effect or explanation
1: *For the first time, the animals are beginning to question the pigs' actions.*	*'Clover … thought she remembered a definite ruling against beds'*	*Because Clover only 'thinks' she remembers, the pigs are able to exploit her poor memory.*
2: *The pigs alter the commandments for their own ends.*		
3: *Squealer verbally manipulates the other animals.*		

❺ Write up **point 1** into a **paragraph** below in your own words. Remember to include what you infer from the evidence, or the writer's effects.

....

....

....

....

....

❻ Now, choose **one** of your **other points** and write it out as another **paragraph** here:

....

....

....

....

....

PROGRESS LOG [tick the correct box] Needs more work ☐ Getting there ☐ Under control ☐

Chapter 7

QUICK TEST

1. Which of the following are the main methods Napoleon uses to rule Animal Farm? **Tick** all those that are correct.
   - a) Violence and intimidation ☐
   - b) Ruthlessness and determination ☐
   - c) Team-building and encouragement ☐

2. Which of the following help to make Napoleon seem more important? **Tick** all the answers that are correct.
   - a) He awards himself medals. ☐
   - b) He uses Snowball as a scapegoat for everything that goes wrong on the farm. ☐
   - c) Squealer re-invents the story of the Battle of the Cowshed. ☐

3. Note down **three** of the crimes of which Snowball is accused in Chapter 7:
   - a) ……………………
   - b) ……………………
   - c) ……………………

THINKING MORE DEEPLY

4. Write **one** or **two sentences** in response to each of these questions:
   - a) How has Squealer changed the story of the Battle of the Cowshed?

     ……………………

     ……………………

     ……………………

     ……………………
   - b) Following the executions, the animals sing 'Beasts of England' 'mournfully' (p. 55). What thoughts and feelings does their singing express?

     ……………………

     ……………………

     ……………………

     ……………………
   - c) Why do you think the pigs decide to ban the singing of 'Beasts of England'?

     ……………………

     ……………………

     ……………………

EXAM PREPARATION: WRITING ABOUT HOW ANIMAL FARM IS RUN A01

Reread the section of Chapter 7 from *'Four days later, in the late afternoon'* (p. 51) to *'unknown there since the expulsion of Jones.'* (p. 53).

Question: What are your impressions of the way in which Animal Farm is now being run?

Think about:

- How Napoleon is presented
- The impact of the executions on the animals and on the reader

⑤ Complete this table:

Point/detail	Evidence	Effect or explanation
1: *Napoleon presents himself as heroic.*	*'wearing both his medals … with his nine huge dogs frisking round him.'*	*Napoleon wears the medals he has awarded himself to win the animals' respect and uses his guard dogs to encourage their fear.*
2: *The number of deaths and their description is horrific.*		
3: *Orwell compares the rule of Napoleon and the rule of Jones.*		

⑥ Write up **point 1** into a **paragraph** below in your own words. Remember to include what you infer from the evidence, or the writer's effects.

....

....

....

....

....

⑦ Now, choose **one** of your **other points** and write it out as another **paragraph** here:

....

....

....

....

....

PROGRESS LOG [tick the correct box] Needs more work ☐ Getting there ☐ Under control ☐

Chapter 8

QUICK TEST

❶ Which of the Seven Commandments have the pigs broken or changed by the end of the chapter? **Tick** any that are correct.

a) Whatever goes upon two legs is an enemy. ☐
b) Whatever goes upon four legs, or has wings, is a friend. ☐
c) No animal shall wear clothes. ☐
d) No animal shall sleep in a bed. ☐
e) No animal shall drink alcohol. ☐
f) No animal shall kill any other animal. ☐
g) All animals are equal. ☐

❷ Look at these 'facts' that the pigs tell the other animals. For each one, decide whether it is **True [T]**, **False [F]** or whether there is **Not Enough Evidence [NEE]** to decide:

a) The farm has produced more food than ever. [T] [F] [NEE]
b) Three hens have confessed to plotting Napoleon's murder. [T] [F] [NEE]
c) Three hens were executed. [T] [F] [NEE]
d) Frederick mistreats his animals and plans to take over Animal Farm. [T] [F] [NEE]
e) Snowball was never awarded the Animal Hero medal. [T] [F] [NEE]
f) The animals won a great victory in the Battle of the Windmill. [T] [F] [NEE]

THINKING MORE DEEPLY

❸ Write **one** or **two sentences** in response to each of these questions:

a) How does Napoleon make himself appear even more important in this chapter?

....

....

....

....

....

b) Why is the fact that Napoleon is seen 'wearing an old bowler hat of Mr Jones's' (p. 67) significant?

....

....

....

....

....

EXAM PREPARATION: WRITING ABOUT THE PRESENTATION OF THE PIGS A01

Reread the section of Chapter 8 from *'It was a few days later'* (p. 67) to the end of the chapter.

Question: How are the pigs presented in this passage?

Think about:

- What the pigs do
- Why they do this

❹ Complete this table:

Point/detail	Evidence	Effect or explanation
1: *The pigs drink Jones's alcohol in the farmhouse.*	*'Napoleon was distinctly seen to … gallop rapidly round the yard'*	*The pigs' privilege and disregard for the commandments is growing.*
2: *The pigs plough up the retirement pasture to grow barley for brewing alcohol.*		
3: *Orwell strongly implies that Squealer has altered the Commandments.*		

❺ Write up **point 1** into a **paragraph** below in your own words. Remember to include what you infer from the evidence, or the writer's effects.

❻ Now, choose **one** of your **other points** and write it out as another **paragraph** here:

PROGRESS LOG [tick the correct box] Needs more work ☐ Getting there ☐ Under control ☐

Chapter 9

QUICK TEST

❶ How many animals on Animal Farm have retired with a pension so far? **Tick** the correct answer.

a) Several ☐

b) One ☐

c) None ☐

❷ What is the real reason for Boxer being sent away from the farm? **Tick** all the correct answers.

a) He is unwell and has to go to hospital. ☐

b) He can no longer work and so is useless. ☐

c) He is approaching the age of retirement. ☐

THINKING MORE DEEPLY

❸ Write **one** or **two sentences** in response to each of these questions:

a) A new rule is introduced requiring animals to stand aside if they meet a pig. What does this suggest about the relationship between the pigs and the other animals?

....

....

....

....

b) Look at the paragraph beginning 'The farm had had a fairly successful year' on page 71. Who benefits from, and who suffers as a result of, the ways in which the pigs make money?

....

....

....

....

c) The animals think that when Jones ran the farm 'they had been slaves and now they were free.' (p. 70) Do you agree? Explain why.

....

....

....

....

....

EXAM PREPARATION: WRITING ABOUT BOXER'S DEATH

A01

Reread the section of Chapter 9 from *'For the next two days Boxer remained in his stall.'* (p. 75) to *'They are taking Boxer to the knacker's!'* (p. 77).

Question: How do you respond to the ways in which Orwell presents Boxer's death?

Think about:

- What Boxer says
- How the animals work out what is happening

❹ Complete this table:

Point/detail	Evidence	Effect or explanation
1: *Boxer has no idea of the fate that awaits him.*	*'he looked forward to the peaceful days … in the corner of the big pasture.'*	*The reader already suspects that retirement is another of the pigs' lies and so tension builds as we wait to see what happens to Boxer.*
2: *The animals have no idea where Boxer is going.*		
3: *Only Benjamin sees what is happening.*		

❺ Write up **point 1** into a **paragraph** below in your own words. Remember to include what you infer from the evidence, or the writer's effects.

....

....

....

....

....

❻ Now, choose **one** of your **other points** and write it out as another **paragraph** here:

....

....

....

....

....

PROGRESS LOG [tick the correct box] Needs more work ☐ Getting there ☐ Under control ☐

Chapter 10

QUICK TEST

❶ Which of the following does Pilkington admire about Animal Farm? (pp. 86–7) **Tick** all the correct answers.

a) Its up-to-date methods of farming ☐

b) Its discipline and orderliness ☐

c) The fact that the animals work harder and eat less than those on any other farm in the country ☐

d) The beer that they brew ☐

❷ What final changes does Napoleon announce at dinner with the other farmers which symbolically confirm that Major's vision no longer has a place on the farm? **Tick** all the correct answers.

a) The farm's name will revert to 'Manor Farm'. ☐

b) A statue of Mr Jones will be placed in the farmyard. ☐

c) A boar's skull has been buried. ☐

d) Any animal who mentions Major will be punished. ☐

THINKING MORE DEEPLY

❸ Write **one** or **two sentences** in response to each of these questions:

a) Squealer says that the pigs work hard all day on 'things called "files", "reports", "minutes" and "memoranda"' (p. 81). What does this suggest about the role of the pigs on the farm?

....

....

....

b) How do you think Orwell wants the reader to respond to the image of pigs on two legs, wearing clothes and smoking pipes?

....

....

....

c) What is the significance of the final sentence of Chapter 10?

....

....

....

....

EXAM PREPARATION: WRITING ABOUT THE ANIMALS' LIVES

A02

Reread the section of Chapter 10 from *'As for the others, their life, so far as they knew'* (p. 81) to *'All animals were equal.'* (p. 83).

Question: In this passage, how does Orwell suggest that nothing has improved or ever will improve for the animals of Animal Farm?

Think about:

- The lives of the animals
- What the animals think and believe

❹ Complete this table:

Point/detail	Evidence	Effect or explanation
1: *Little has changed in the lives of the animals.*	*'They were generally hungry … they laboured in the fields'*	*It is now several years later; nothing has changed and it seems unlikely things will ever change.*
2: *The animals still believe in the principles of Animalism.*		
3: *The animals believe everything they are told, even though they have no evidence to support it.*		

❺ Write up **point 1** into a **paragraph** below in your own words. Remember to include what you infer from the evidence, or the writer's effects.

....

....

....

....

....

❻ Now, choose **one** of your **other points** and write it out as another **paragraph** here:

....

....

....

....

....

PROGRESS LOG [tick the correct box] Needs more work ☐ Getting there ☐ Under control ☐

Practice task

❶ First, **read** this **exam-style** task:

> Read the section from *'November came, with raging south-west winds'* (Ch. 6, p. 45) to the end of Chapter 6.
>
> Question: In this passage, what methods does Orwell use to present characters and events? Refer closely to the passage in your answer.

❷ Begin by circling the **key words** in the **question** above.

❸ Now complete this table, noting down **three or four key points** with **evidence** and the **effect created**.

Point	Evidence/quotation	Meaning or effect

❹ **Draft your response.** Use the space below for your first paragraph(s) and then continue onto a sheet of paper.

Start: *In this extract, Orwell describes the animals' reaction to the destruction of the windmill by a storm …* ....

....

....

....

....

....

....

....

....

PROGRESS LOG [tick the correct box] Needs more work ☐ Getting there ☐ Under control ☐

PART THREE: CHARACTERS

Who's who?

Look at these drawings and **complete** the **name(s)** of each of the characters shown.

Major

❶ Look at this bank of **adjectives** describing Major. Circle the ones you think best **describe** him.

imaginative	*respected*	*intelligent*
visionary	*quick-tempered*	*influential*
impulsive	*disillusioned*	*wise*

❷ Now add the **page reference** from your copy of the book next to each circle, showing where evidence can be found to **support** the **adjective**.

❸ Look at the way in which Major's ideas are introduced and developed:

Major makes a speech.	→	The pigs announce the Seven Commandments.	→	The Seven Commandments are reduced to a single, simplified commandment.	→	The single commandment is rewritten.

Write **two sentences** summarising what happens to Major's ideas.

....

....

....

....

❹ Look at these facts about the philosopher Karl Marx. Write a sentence explaining how each of Marx's ideas is similar to one of Major's ideas.

a) *Marx wrote the Communist Manifesto. Major …* ....

....

....

b) *Marx argued that the rich have always exploited working people. Major …* ....

....

....

c) *Marx argued that the only solution was for working people to stage a revolution and seize power from the rich. Major …* ....

....

....

PROGRESS LOG [tick the correct box] Needs more work ☐ Getting there ☐ Under control ☐

Snowball

❶ The pigs spread many lies about Snowball in order to to discredit him. Which of these statements are **TRUE** and which are **FALSE**? Write **'T'** or **'F'** in the boxes:

a) It was Snowball's idea to build a windmill. ☐

b) Snowball destroyed the windmill. ☐

c) Snowball was a spy for Mr Jones. ☐

d) Snowball led the animals to victory in the Battle of the Cowshed. ☐

❷ **Complete** these quotations, which either describe Snowball or are said by him.

a) 'Snowball was a more .... pig than Napoleon, quicker in .... and more ....'

b) 'Then Snowball (for it was Snowball who was best at ....) took a brush between the two knuckles of his trotter, painted out .... from the top bar of the gate and in its place painted ....'

c) 'The birds did not understand Snowball's .... , but they accepted his ....'

d) 'The only .... human being is a .... one.'

e) 'One of them all but closed his .... on Snowball's ...., but Snowball whisked it free just in time.'

❸ Write a **paragraph** explaining how Orwell presents Snowball. Try to use one of the **quotations** above, or choose a different one to support what you say:

I believe Orwell wants to present Snowball as … ....

....

....

....

....

....

....

....

....

....

....

....

....

PROGRESS LOG [tick the correct box] Needs more work ☐ Getting there ☐ Under control ☐

Napoleon

❶ Each of the character traits in the table below could be applied to Napoleon. Working from **memory** add points in the story when you think these are shown, then find at least one **quotation** to back up your ideas.

Quality	Moment/s in story	Quotation
a) Arrogant		
b) Ruthless		
c) Selfish		
e) Tyrannical		

❷ **Circle** any of the following adjectives that you would add to the list.

dictatorial *hypocritical* *principled* *lazy*

manipulative *devious* *cowardly*

impulsive *intelligent* *shrewd* *disloyal*

❸ Write **two sentences** in response to each of these questions. You could use some of the **adjectives** above in your answers.

a) Why do you think Napoleon wants to be the leader of Animal Farm?

....

....

....

b) Which of Napoleon's actions or decisions do you find most shocking or disturbing?

....

....

....

PROGRESS LOG [tick the correct box] Needs more work ☐ Getting there ☐ Under control ☐

Squealer

1. Look at these statements about Squealer. For each one, decide whether it is **True [T]**, **False [F]** or whether there is **Not Enough Evidence [NEE]** to decide:

   a) Squealer's main role on the farm is to spread Napoleon's propaganda. [T] [F] [NEE]

   b) Squealer does not fight in the Battle of the Windmill. [T] [F] [NEE]

   c) Squealer enjoys the power and privileges of being a pig. [T] [F] [NEE]

   d) Squealer cries when he tells the animals of Boxer's death. [T] [F] [NEE]

   e) Squealer is deeply upset by the death of Boxer. [T] [F] [NEE]

2. Squealer uses the following **tactics** to control and manipulate the other animals. Add an **example** of Squealer using each tactic to the table. The first one has been done to help you.

Tactic	Example
Lies	*'Many of us actually dislike milk and apples.'*
Fear	
False facts and statistics	
False arguments	
Rhetorical questions	
Repetition	

3. Using your **own judgement**, put a mark along this line to show **Orwell's overall presentation** of Squealer.

Not at all sympathetic ① — A little sympathetic ② — Quite sympathetic ③ — Very sympathetic ④

PROGRESS LOG [tick the correct box] Needs more work ☐ Getting there ☐ Under control ☐

Boxer

1. **Complete** these **statements** about Boxer:

   a) *Boxer is enormously strong but not very intelligent. For example …*

   b) *Orwell emphasises how hard Boxer works when …*

   c) *Boxer's bravery is shown when he …*

   d) *Boxer's honesty and sensitivity are shown in the Battle of the Cowshed when he …*

2. Look at these **quotations** about Boxer. What do they reveal about his **character** and/or the way Orwell uses his character to present his **ideas**?

Quotation	Effect or explanation
Major warns him, 'the very day that those great muscles of yours lose their power, Jones will send you to the knacker' (Ch. 1, p. 5)	
Boxer's answer to everything is 'I will work harder.' (Ch. 3, p. 17)	
Boxer believes that 'Napoleon is always right.' (Ch. 3, p. 35)	
After he is injured, 'he looked forward to the peaceful days that he would spend in the corner of the big pasture.' (Ch. 9, p. 76)	

3. On a separate sheet of paper, write a **paragraph** explaining how Orwell presents Boxer. Try to use one of the **quotations** above, or one that you have chosen, to support what you say. Start:

   I believe Orwell wants to present Boxer as …

PROGRESS LOG [tick the correct box] Needs more work ☐ Getting there ☐ Under control ☐

Clover

❶ **Complete** these **statements** about Clover:

a) *The reader knows that Clover is slightly more intelligent than Boxer because she …*

b) *After the executions, Clover cannot express her thoughts and feelings so she …*

c) *When Boxer is hurt, she …*

d) *When Clover sees Squealer walking on two legs, she …*

❷ Each of these qualities could be applied to Clover. Working from **memory** add points in the story when you think these are shown, then find at least one **quotation** to back up your ideas.

Quality	Moment/s in story	Quotation
a) Kind and caring		
b) Unintelligent		
c) Questioning		
d) Loyal		

❸ Using your **own judgement**, put a mark along this line to show **Orwell's overall presentation** of Clover.

❶	❷	❸	❹
Not at all sympathetic	A little sympathetic	Quite sympathetic	Very sympathetic

PROGRESS LOG [tick the correct box] Needs more work ☐ Getting there ☐ Under control ☐

Benjamin

❶ **Draw a line** linking each piece of evidence to the point which it supports.

Point
1 Benjamin is cynical and miserable.
2 Benjamin is kind and caring.
3 Benjamin seems to understand everything but does nothing.

Evidence
A When Benjamin tells the others that Boxer is being taken away: 'it was the first time that anyone had ever seen him gallop' (Ch. 9, p. 76)
B When Frederick's men are about to blow up the windmill: '"I thought so," he said. "Do you not see what they are doing?"' (Ch. 8, p. 64)
C 'Benjamin professed … to know that things had never been, nor ever could be much better or much worse' (Ch. 10, p. 82)

❷ Write **one or two sentences**, explaining how each quotation supports the point to which you have linked it.

A ....

B ....

C ....

❸ Write **a paragraph** explaining why you think Benjamin will not say or do anything to oppose the changes to Animal Farm and to help the animals.

....

PROGRESS LOG [tick the correct box] Needs more work ☐ Getting there ☐ Under control ☐

Mr Jones

❶ Look at these points about Mr Jones. **Complete** the table, explaining how you think Orwell wants the reader to respond to the character of Jones and supporting your ideas with evidence from the text.

Point	Evidence	How Orwell wants the reader to respond
1: *Mr Jones does not look after his animals or his farm very well and is driven from his farm by his own animals.*		
2: *Mr Jones and the neighbouring farmers try to recapture the farm but are defeated.*		
3: *Mr Jones sometimes used to mix milk in with the hens' mash.*		
4: *Orwell tells us what has happened to Mr Jones.*		

❷ If we read *Animal Farm* as a **political allegory**, Mr Jones could be said to represent **Tsar Nicholas II**, the ruler of Russia, who was overthrown in the Russian Revolution of 1917. Complete this **gap-fill** paragraph about the Tsar to show what you can infer about Orwell's opinion of the way Nicholas II ruled Russia:

Orwell seems to suggest that the Tsar's rule of Russia was .... and .... : that he took no interest in the .... of his people and no interest in the .... of those people whose job it was to look after them.

❸ Using your **own judgement**, put a mark along this line to show Orwell's overall presentation of Mr Jones.

Not at all sympathetic ❶ — A little sympathetic ❷ — Quite sympathetic ❸ — Very sympathetic ❹

PROGRESS LOG [tick the correct box] Needs more work ☐ Getting there ☐ Under control ☐

Moses

❶ Complete these **quotations** describing Moses. Use the words and phrases in the box below:

a) 'The pigs had an even harder struggle to counteract the .... put about by Moses.'

b) 'Moses, who was Mr Jones's .... , was a .... and a ....'

c) 'The animals hated Moses because he told .... and did not ....'

lies	*especial pet*	*spy*	*tale-bearer*
tales		*work*	

❷ Write **two sentences** in response to each of these questions:

a) Look again at Moses's description of Sugarcandy Mountain on page 10. Why do you think some of the animals want to believe Moses?

....

....

....

....

....

b) Why do you think some of the animals don't believe Moses?

....

....

....

....

....

c) Moses follows Mr and Mrs Jones when they flee the farm and does not return until Chapter 9. Why do you think the pigs allow Moses to return, and give him 'an allowance of a gill of beer a day' (Ch. 9, p. 74)?

....

....

....

....

....

PROGRESS LOG [tick the correct box] Needs more work ☐ Getting there ☐ Under control ☐

Minor characters

MOLLIE

1 Which of the following does Mollie **Like [L]**, **Dislike [D]** or **Neither like nor dislike [N]**?

a) Sugar [L] [D] [N]

b) Ribbons [L] [D] [N]

c) Hard work [L] [D] [N]

d) Animalism [L] [D] [N]

2 If we read *Animal Farm* as a **political allegory**, Mollie could be said to represent the privileged **Russian middle class**, many of whom fled Russia after the 1917 revolution. Complete this **gap-fill** about the Russian middle class to show what you can infer about Orwell's opinion of them.

Orwell seems to be suggesting that the Russian middle class were

.... and .... and .... in the lives of

those worse off than themselves.

JESSIE, PINCHER AND BLUEBELL

3 Orwell tells us very little about these three farm dogs: they chase the rats in Chapter 1 and give birth to nine puppies in Chapter 3. Write **one** or **two sentences** explaining why you think Orwell included them.

....

....

....

NAPOLEON'S DOGS

4 **Complete** the table below, writing **one** or **two sentences** explaining what each quotation suggests about the dogs and/or the pigs.

Quotation	Explanation
'there was a terrible baying sound outside, and nine enormous dogs wearing brass-studded collars came bounding into the barn. They dashed straight for Snowball' (Ch. 5, p. 33)	
'Squealer spoke so persuasively, and the three dogs who happened to be with him growled so threateningly, that they accepted his explanation without further questions' (Ch. 5, p. 37)	
'Napoleon himself was not seen in public as often as once a fortnight. When he did appear, he was attended … by his retinue of dogs' (Ch. 8, p. 57)	

THE SHEEP

❺ What role do the sheep play in *Animal Farm*? Why do you think Orwell gave this role to the sheep? Write **one** or **two sentences** explaining your ideas.

....

....

....

....

MURIEL

❻ What do these quotations suggest about the character of Muriel? **Complete** the table, adding **one** or **two adjectives** for each quotation.

Evidence	Adjectives
• In the Battle of the Cowshed, she 'rushed forward and prodded and butted the men from every side' (Ch. 4, p. 25)	
• 'Muriel … could read somewhat better than the dogs, and sometimes used to read to the others in the evenings from scraps of newspaper which she found on the rubbish heap.' (Ch. 3, p. 20)	
• Muriel reads the Sixth Commandment to Clover when Clover thinks it may have been changed but Muriel does not comment on the change.	

MR FREDERICK, MR PILKINGTON AND MR WHYMPER

❼ **Tick** the correct character's name:

a) Who owns Foxwood Farm?
Mr Frederick ☐ Mr Pilkington ☐ Mr Whymper ☐

b) Who owns Pinchfield Farm and blows up the windmill?
Mr Frederick ☐ Mr Pilkington ☐ Mr Whymper ☐

c) Who is the solicitor Napoleon employs to trade with humans?
Mr Frederick ☐ Mr Pilkington ☐ Mr Whymper ☐

❽ Which of the following adjectives could you use to describe these men? Add an initial (**F, P** and/or **W**) below to each adjective to indicate who it could be used to describe.

devious	dishonest	ruthless	cruel	aggressive	greedy
....	....	....	....	....	....

PROGRESS LOG [tick the correct box] Needs more work ☐ Getting there ☐ Under control ☐

Practice task

1. First, **read** this **exam-style** task:

Read the section from *'Three days later it was announced'* (Ch. 9, p. 78) to *'he had died happy'* (Ch. 9, p. 79).

Question: In this passage, what methods does Orwell use to present the character of Squealer? Refer closely to the passage in your answer.

2. Begin by circling the **key words** in the **question** above.

3. Now complete this table, noting down **three or four key points** with **evidence** and the **effect created**.

Point	Evidence/quotation	Meaning or effect

4. **Draft your response**. Use the space below for your first paragraph(s) and then continue onto a sheet of paper.

Start: *In this extract, Orwell presents …* ....

....

....

....

....

....

....

....

....

....

PROGRESS LOG [tick the correct box] Needs more work ☐ Getting there ☐ Under control ☐

PART FOUR: THEMES, CONTEXTS AND SETTINGS

Themes

QUICK TEST ✓

❶ Check your understanding of some of the **key ideas** in *Animal Farm*. **Draw** a **line** linking each idea to the correct definition:

a) Equality	1 Biased or misleading information used to support a point of view
b) Propaganda	2 Dishonest activity by those in power
c) Corruption	3 The cruel misuse of power
d) Tyranny	4 Having the same rights and opportunities as others

❷ Circle the themes you think are most **relevant** to *Animal Farm:*

power *equality* *dreams* *propaganda*

society *corruption* *language*

violence *revolution* *education* *tyranny*

greed *cruelty*

❸ Choose what you consider to be the three most important themes of the text and **complete** the following **statements**:

a) Theme 1: *Two incidents where Orwell explores the theme of* ....

are when ....

and when ....

b) Theme 2: *Two incidents where Orwell explores the theme of* ....

are when ....

and when ....

c) Theme 3: *Two incidents where Orwell explores the theme of* ....

are when ....

and when ....

THINKING MORE DEEPLY ?

4. Look again at your answers to Question 3. What can you infer about **Orwell's views** on the three key themes you wrote about? Write **one** or **two sentences** about each one.

   a) *In his presentation of the theme of* ....
   Orwell seems to be suggesting that ....

   b) *In his presentation of the theme of* ....
   Orwell seems to be suggesting that ....

   c) *In his presentation of the theme of* ....
   Orwell seems to be suggesting that ....

5. Which themes does Orwell explore through each of the following characters? Write **one or more** of your chosen **themes** next to each character. Add **one** or **two sentences** explaining **how** Orwell uses that character to explore the theme(s):

   a) *Orwell uses the character of Major to explore the theme(s) of* ....
   For example, ...

   b) *Orwell uses the character of Snowball to explore the theme(s) of* ....
   For example, ...

   c) *Orwell uses the character of Napoleon to explore the theme(s) of* ....
   For example, ...

d) *Orwell uses the character of Squealer to explore the theme(s) of* ....
For example, ...

....

....

....

e) *Orwell uses the character of Boxer to explore the theme(s) of* ....
For example, ...

....

....

....

f) *Orwell uses the character of Mr Jones to explore the theme(s) of* ....
For example, ...

....

....

....

⑥ It is important that you use **quotations** to explore themes. Read this quotation and then:

- identify the theme that it represents.
- add further annotations, explaining how specific words or phrases in the quotation relate to this theme.

Theme: ....

Squealer uses this term to make the animals feel included

'"Surely, comrades," cried Squealer almost pleadingly… "surely there is no one among you who wants to see Jones come back?"' (Ch. 3, p. 22)

EXAM PREPARATION: WRITING ABOUT POWER AND CORRUPTION

A01

Question: In the text as a whole, how does Orwell present the theme of power and corruption?

Think about:

- The animals' ideas when they first took over the farm
- How the pigs behave and what they say as the text develops

7 Complete this table:

Point/detail	Evidence	Effect or explanation
1: *The pigs write the Seven Commandments on the barn wall but then go on to change them to suit their own corrupt ends.*	*'All animals are equal but some animals are more equal than others.'*	*The pigs use their intelligence to control the less intelligent animals, turning the most important commandment into nonsense.*
2: *The pigs use fear and violence to control the animals.*		
3: *Squealer uses the power of language to exploit the animals' fears and manipulate the truth.*		

8 Write up **point 1** into a **paragraph** below in your own words. Remember to include what you infer from the evidence, or the writer's effects.

9 Now, choose **one** of your **other points** and write it out as another **paragraph** here:

PROGRESS LOG [tick the correct box] Needs more work ☐ Getting there ☐ Under control ☐

Contexts

QUICK TEST

1. In which year was *Animal Farm* written? **Tick** the correct answer.
   a) 1917 ☐
   b) 1933 ☐
   c) 1945 ☐
2. In which year did the Russian Revolution take place? **Tick** the correct answer.
   a) 1917 ☐
   b) 1933 ☐
   c) 1945 ☐
3. When did the Second World War start and finish? **Tick** the correct answer.
   a) 1914–18 ☐
   b) 1933–37 ☐
   c) 1939–45 ☐
4. What is the correct definition of 'totalitarianism'? **Tick** the correct answer.
   a) A political system in which individuals have total freedom ☐
   b) A political system in which rulers have total power and control over society and individuals have no freedom ☐
   c) A political system in which people vote for their leaders ☐
   d) A political system in which there are no leaders ☐
5. Which of the following countries were ruled by totalitarian leaders when Orwell was writing *Animal Farm*? **Tick** all of the correct answers.
   a) Great Britain ☐
   b) The Soviet Union ☐
   c) Italy ☐
   d) The United States of America ☐
   e) Germany ☐

George Orwell

6 Draw lines to show which countries were ruled by which kind of leader or government when Orwell was writing *Animal Farm*.

Great Britain	
The Soviet Union	democratic
Italy	fascist
The United States of America	Communist
Germany	

THINKING MORE DEEPLY ?

Animal Farm can be read as a political allegory symbolising the hopes of Communism, the Russian Revolution and the transformation of Russia into the Union of Soviet Socialist Republics (USSR).

7 Using the contextual information on page 52, write **one** or **two sentences** in response to each of these questions:

a) At the end of *Animal Farm*, the animals peer into the farmhouse and cannot tell the difference between the pigs and the humans. What do you think Orwell is suggesting about the political situation at the time he was writing *Animal Farm*?

....

....

....

....

....

b) Which other political leaders of 1945 might be said to be reflected in the character and actions of Napoleon? Write one or two sentences explaining your ideas.

....

....

....

....

....

c) Orwell was a socialist: he believed that all people should be as equal as possible. Why do you think in *Animal Farm* he chose to depict the failure of his beliefs?

....

....

....

....

....

8 Compare the events from *Animal Farm* and events in history below. **Draw lines** linking those that seem to have significant similarities.

1 In 1848 Karl Marx published the *Communist Manifesto* in which he argued that the poor could only improve their lives through revolution.

2 In 1917 Lenin and the Communist Party seized power from, and later executed, the Tsar of Russia and his family.

3 Russia's name was changed to the Union of Soviet Socialist Republics, or the Soviet Union.

4 Stalin and Trotsky tried to apply Marx's ideas in Russia.

5 Stalin forced Trotsky out of power and out of the Soviet Union.

6 Stalin became the sole ruler of Russia, using fear and propaganda to maintain power.

7 During the Second World War, Stalin negotiated with both Britain and Hitler's Germany to protect the Soviet Union from attack.

8 In 1941, Germany invaded Russia.

9 Britain, the Soviet Union and the USA met at the Tehran Conference in 1943 and negotiated a fragile alliance which soon collapsed.

A Snowball and Napoleon summarise Major's ideas in the Seven Commandments. (Chapter 2)

B Napoleon, Frederick and Pilkington eat dinner together but soon argue. (Chapter 10)

C Major explains his dream and his vision for the future. (Chapter 1)

D The animals change the name of Manor Farm to Animal Farm. (Chapter 2)

E Snowball flees Animal Farm. (Chapter 5)

F Frederick and his men destroy the windmill. (Chapter 8)

G Napoleon promises to sell a stack of timber to both neighbouring farmers, Mr Pilkington and Mr Frederick.

H Napoleon becomes more and more powerful. (Chapters 5–10)

I Snowball leads the animals' rebellion and Jones is expelled from his farm. (Chapter 2)

Marx

Lenin

Trotsky

Stalin

EXAM PREPARATION: WRITING ABOUT ALLEGORY

A03

Question: George Orwell wrote *Animal Farm* in 1945. What do you learn from the text about Orwell's hopes and fears for society and its rulers at that time?

Think about:

- The allegorical connections between the characters and the USSR
- How Orwell uses allegory to comment on the political situation

⑨ Complete this table:

Point	Evidence/quotation	Meaning or effect
1: *The ideals of the rebellion are soon lost.*	*'when they came back in the evening it was noticed that the milk had disappeared.'*	*Orwell is suggesting that the Marxist ideals of the Russian Revolution were destroyed by its leaders.*
2: *The pigs treat the other animals violently and ruthlessly.*		
3: *By the end, the animals cannot tell the pigs and men apart.*		

⑩ Write up **point 1** into a **paragraph** below in your own words. Remember to include what you infer from the evidence, or the writer's effects.

⑪ Now, choose **one** of your **other points** and write it out as another **paragraph** here:

PROGRESS LOG [tick the correct box] Needs more work ☐ Getting there ☐ Under control ☐

Settings

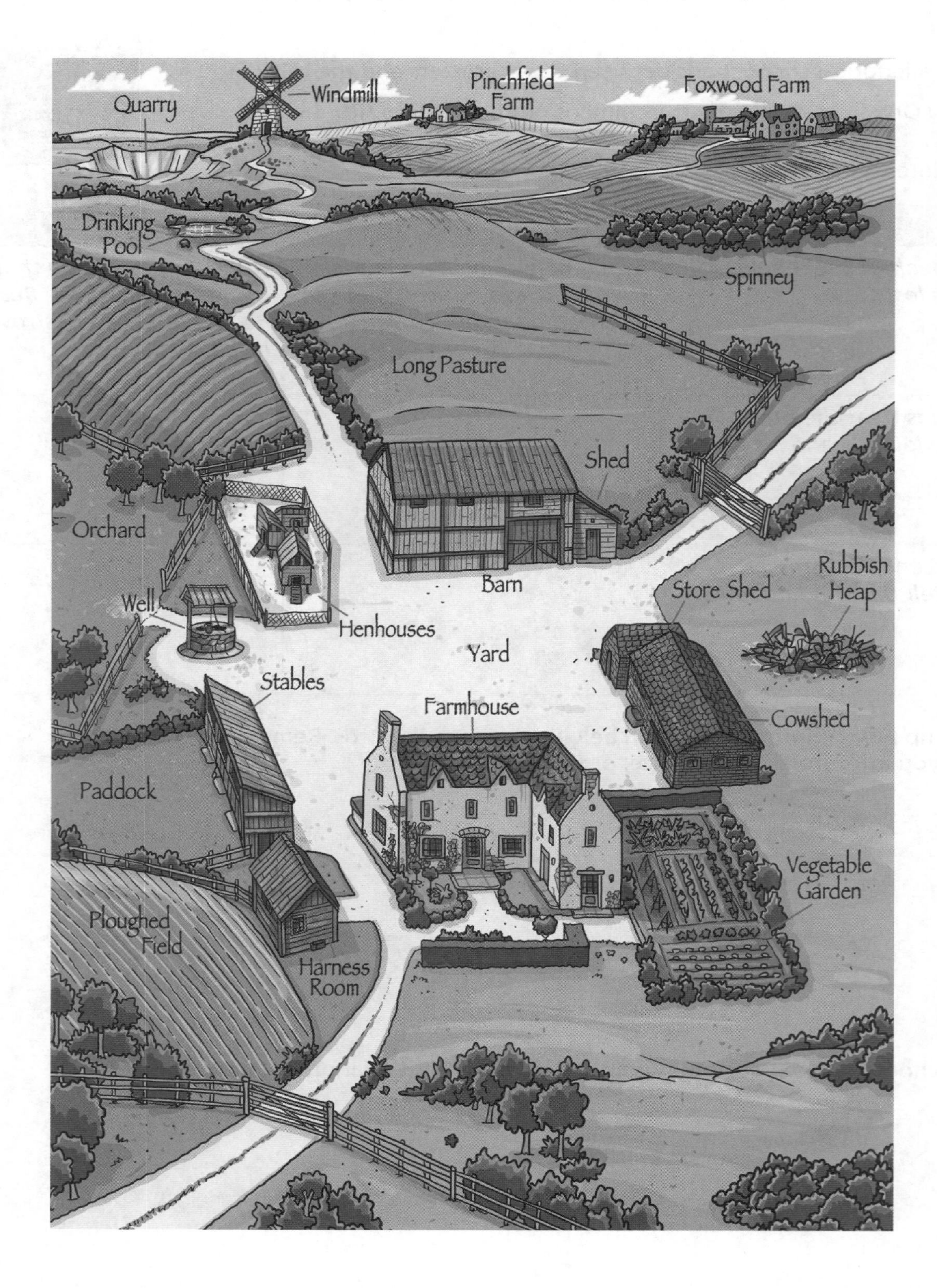
Quarry
Windmill
Pinchfield Farm
Foxwood Farm
Drinking Pool
Spinney
Long Pasture
Shed
Orchard
Barn
Store Shed
Rubbish Heap
Well
Henhouses
Yard
Stables
Farmhouse
Cowshed
Paddock
Vegetable Garden
Ploughed Field
Harness Room

QUICK TEST ✓

1. Add labels (a–k) to the diagram on page 54 to show where:

   a) Mr Frederick lives
   b) Mr Jones lives at the beginning of *Animal Farm*
   c) the mass executions of Chapter 7 take place
   d) Major calls the animals together to hear his dream
   e) Napoleon and the pigs live and work by the end of *Animal Farm*
   f) the pigs set up their first headquarters and learn to read
   g) the windmill is debated and Snowball is chased away
   h) Boxer collapses
   i) the Seven Commandments are painted on the wall
   j) the puppies are kept and educated in a loft
   k) Mr Pilkington lives

THINKING MORE DEEPLY ?

2. Write **one** or **two sentences** in response to each of these questions:

   a) How does Orwell use the setting of the farmhouse to first draw attention to the growing similarities between the pigs and human beings?

   ....

   ....

   ....

   ....

   b) **Compare** these two quotations:

'It was a clear spring evening. The grass and the bursting hedges were gilded by the level rays of the sun. Never had the farm … appeared to the animals so desirable a place.' (Ch. 7, p. 54)	'They were generally hungry, they slept on straw, they drank from the pool, they laboured in the fields; in winter they were troubled by the cold, and in the summer by the flies.' (Ch. 10, p. 81)

How does Orwell use the setting of a farm to highlight the difference between the dream of Animalism and the reality of Animalism? Write **one** to **two sentences** explaining your ideas.

....

....

....

....

PROGRESS LOG [tick the correct box] Needs more work ☐ Getting there ☐ Under control ☐

Practice task

❶ First, **read** this **exam-style** task:

Question: How does George Orwell use the **setting** of **the barn** to highlight the changes that take place on Animal Farm?

❷ Begin by circling the **key words** in the **question** above.

❸ Now complete this table, noting down **three or four key points** with **evidence** and the **effect created**.

Point	Evidence/quotation	Meaning or effect

❹ **Draft your response.** Use the space below for your first paragraph(s) and then continue onto a sheet of paper.

Start: *In 'Animal Farm', Orwell uses the setting of the barn in a number of ways. In Chapter 1 …* ....

....

....

....

....

....

....

....

....

....

PROGRESS LOG [tick the correct box] Needs more work ☐ Getting there ☐ Under control ☐

PART FIVE: FORM, STRUCTURE AND LANGUAGE

Form

QUICK TEST

❶ Orwell gave *Animal Farm* the subtitle 'A Fairy Story'. Look at these common elements of fairy tales. **Tick** all those that feature in *Animal Farm*:

a) Animals that can talk ☐

b) Magical events or transformations ☐

c) Elements of fantasy ☐

d) Violence and cruelty ☐

e) Good characters and evil characters ☐

f) The good are rewarded and the bad are punished ☐

g) A moral: a lesson for the reader to learn ☐

❷ Which **one** of the following best explains why *Animal Farm* can be considered a **satire**?

a) It is like the story 'The Tortoise and the Hare': it uses animals with human qualities to make a moral point. ☐

b) It mocks farmyard animals, making them behave like human beings. ☐

c) It uses exaggeration and irony to expose and ridicule negative qualities and weaknesses. ☐

THINKING MORE DEEPLY

❸ Write **one** or **two sentences** in response to each question:

a) Why do you think Orwell gave *Animal Farm* the subtitle 'A Fairy Story'?

....

....

....

....

....

b) *Animal Farm* can be considered as a satire: how does Orwell's choice of a pig to take the role of a tyrant satirise politicians who seek power?

....

....

....

....

PROGRESS LOG [tick the correct box] Needs more work ☐ Getting there ☐ Under control ☐

Structure

QUICK TEST

❶ The structure of *Animal Farm* could be summed up in the three sentences below. Below each one, write another sentence adding more detail.

a) At first, things go well.

....

....

b) Then, things start to go wrong.

....

....

c) Finally, things go very badly wrong.

....

....

❷ Which of the following does Orwell use to help the reader chart the failure of Major's vision and the descent of Animal Farm into tyranny? **Tick** all the correct answers.

a) Jones is shown to be an incompetent and uncaring farmer. ☐
b) Major describes his vision of the future. ☐
c) The animals take over the farm and the first harvest is a huge success. ☐
d) The animals work together to win a great victory at the Battle of the Cowshed. ☐
e) The pigs break more and more of the Seven Commandments. ☐
f) Napoleon's actions become increasingly violent. ☐
g) The animals cannot tell the difference between the pigs and the humans. ☐

THINKING MORE DEEPLY

❸ Write **one** or **two sentences** in response to the following questions:

a) The structure of *Animal Farm* could be described as circular. In what way does the story end where it began?

....

....

....

b) At what point in the text did you realise that *Animal Farm* was not going to have a happy ending?

....

....

....

EXAM PREPARATION: WRITING ABOUT STRUCTURE

AO2

Question: How does George Orwell use the **structure** of the text to make the reader feel shocked and disappointed by the events he describes in *Animal Farm*?

Think about:

- How the novel begins and ends
- How Orwell shows the ways in which the farm changes throughout the novel

❹ Complete this table:

Point	Evidence/quotation	Meaning or effect
1: *Each of the Seven Commandments is broken and altered.*	*'No animal shall sleep in a bed with sheets'*	*Each change to the commandments signals the pigs' growing corruption.*
2: *At the end, the pigs walk on two legs and wear human clothes.*		
3: *Repeatedly the animals do not question or comment on the pigs' actions.*		

❺ Write up **point 1** into a **paragraph** below in your own words. Remember to include what you infer from the evidence, or the writer's effects.

....

....

....

....

....

....

❻ Now, choose **one** of your **other points** and write it out as another **paragraph** here:

....

....

....

....

....

....

PROGRESS LOG [tick the correct box] Needs more work ☐ Getting there ☐ Under control ☐

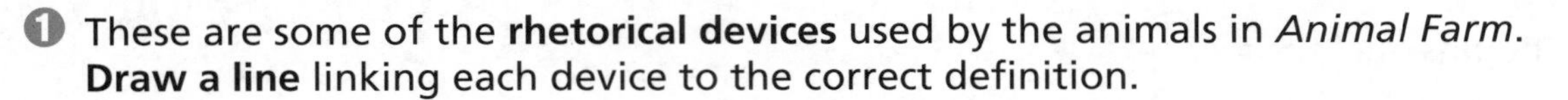

Language

QUICK TEST

1. These are some of the **rhetorical devices** used by the animals in *Animal Farm*. **Draw a line** linking each device to the correct definition.

Rhetorical device	Definition
a) Contrast	A question that needs no answer
b) Rhetorical question	A list of three items, adjectives, ideas, etc.
c) Pattern of three	Language that appeals to the reader or listener's emotions
d) Emotive language	Numerical facts used to prove a point
e) Repetition	Emphasising the difference between two things
f) Statistics	Using the same word, phrase or sentence more than once

2. Orwell uses **irony** and **dramatic irony** to present his ideas. **Draw lines** to show which form of irony matches which definition.

Irony	Saying one thing but meaning exactly the opposite
Dramatic irony	When the writer provides enough information for the reader to infer or understand more than the characters in a text understand

THINKING MORE DEEPLY

3. Look at the section of Major's speech from 'Now, Comrades what is the nature of this life of ours?' on page 3 to 'abolished for ever' on page 4. **Add an example** of each **rhetorical device** from the extract to the table below:

Rhetorical device	Example
Contrast	
Rhetorical question	
Pattern of three	
Emotive language	
Repetition	
Statistics	

❹ Look at Squealer's speech in Chapter 3 (page 22) where he explains why the pigs deserve the milk and apples. **Add** some of the **rhetorical devices** he uses to the table below, and an **example** of each one.

Rhetorical device	Example

❺ Which of the following do you think is most effective in persuading the animals to accept the pigs' decisions?

a) Language ☐

b) Violence ☐

c) Fear ☐

❻ Write **one** or **two sentences** explaining your answer to Question 5.

....

....

....

....

❼ Look at these events and then indicate whether each one is an example of **Irony [I]** or **Dramatic Irony [DI]** or **Neither [N]**:

a) Major warns Boxer that Jones will have him slaughtered when he can no longer work. [I] [DI] [N]

b) The more the animals suffer, the more they are told, and believe, that things are getting better. [I] [DI] [N]

c) It is not made clear where the pigs get the money to buy whisky just after Boxer is taken away. [I] [DI] [N]

d) The animals cannot understand why Squealer is lying on the ground next to a ladder with a pot of paint. [I] [DI] [N]

e) The exhausted, starving animals must celebrate the success of Animal Farm in a Spontaneous Demonstration. [I] [DI] [N]

⑧ Look at this quotation from Chapter 9, describing the moment Boxer is taken away:

> 'Boxer's face did not reappear at the window. Too late, someone thought of racing ahead and shutting the five-barred gate; but in another moment the van was through it and rapidly disappearing down the road. Boxer was never seen again.' (p. 77)

a) Which of these language features has Orwell used in the quotation above?

A short sentence ☐

Little or no description ☐

The other animal's thoughts and feelings about what is happening ☐

Simple vocabulary ☐

b) Highlight an example of all those he has used in the quotation and label them.

c) How does the use, or absence, of each of these features contribute to making this one of the most dramatic and emotive moments in the text? Write **one** or **two sentences**, explaining your ideas.

....

....

....

....

⑨ Choose a quotation of two or three sentences from another dramatic moment in the text. You could choose:

- The expulsion of Snowball in Chapter 5 (page 33)
- The executions in Chapter 7 (page 53)
- The destruction of the windmill in Chapter 8 (page 64)

a) Write your chosen quotation below:

....

....

....

b) Now write **one** or **two sentences** in your own words about how Orwell has used language techniques to make this moment dramatic or tense.

....

....

....

....

....

PROGRESS LOG [tick the correct box] Needs more work ☐ Getting there ☐ Under control ☐

Practice task

1. First, **read** this **exam-style** task:

Question: How does George Orwell use language techniques to show the impact of the pigs' rule on the other animals' lives?

2. Begin by circling the **key words** in the **question** above.

3. Now complete this table, noting down **three or four key points** with **evidence** and the **effect created**.

Point	Evidence/quotation	Meaning or effect

4. **Draft your response**. Use the space below for your first paragraph(s) and then continue onto a sheet of paper.

Start: *Orwell uses a variety of language techniques to show the impact of …* ....

PROGRESS LOG [tick the correct box] Needs more work ☐ Getting there ☐ Under control ☐

PART SIX: PROGRESS BOOSTER

Writing skills

❶ How well can you express your ideas about *Animal Farm*? Look at this grid and tick the level you think you are currently at:

Level	How you respond	What your spelling, punctuation and grammar are like	Tick
Higher	• You analyse the effect of specific words and phrases very closely (i.e. 'zooming in' on them and exploring their meaning). • You select quotations very carefully and you embed them fluently in your sentences. • You are persuasive and convincing in the points you make, often coming up with original ideas.	• You use a wide range of specialist terms (words like 'imagery'), and accurate punctuation, spelling and grammar.	
Mid	• You analyse some parts of the text closely, but not all the time. • You support what you say with evidence and quotations, but sometimes your writing could be more fluent to read. • You make relevant comments on the text.	• You use a good range of specialist terms, and generally accurate punctuation, spelling and grammar.	
Lower	• You comment on some words and phrases but often you do not develop your ideas. • You sometimes use quotations to back up what you say but they are not always well chosen. • You mention the effect of certain words and phrases but these are not always relevant to the task.	• You do not have a very wide range of specialist terms, but you use reasonably accurate spelling, punctuation and grammar.	

SELECTING AND USING QUOTATIONS

❷ Read these two samples from students' responses to a question about how Snowball is presented. Decide which of the three levels they fit best, i.e. **lower (L)**, **mid (M)** or **higher (H)**.

Student A: *Snowball uses complicated language when he explains things to the other animals. For example, he explains that birds are not two-legged animals because '"A bird's wing, comrades," he said, "is an organ of propulsion and not of manipulation."' He is using long words to confuse and mislead the other animals.*

Level ? ☐ Why? ....

....

Student B: *Orwell suggests how intelligent Snowball is when he explains that a bird's wing is 'an organ of propulsion and not of manipulation'. While the animals accept the explanation it is not clear that they understand Snowball's complex language choices. Orwell seems to be implying that Snowball's intentions are good but his ability to communicate with less intelligent creatures is ineffective and he is therefore perhaps not as manipulative as some of the other pigs.*

Level ? ☐ Why? ....

....

ZOOMING IN – YOUR TURN!

Here is the first part of another student response. The student has picked a good quotation but he hasn't 'zoomed in' on any particular words or phrases:

Old Major tells the animals that 'the very instant that our usefulness has come to an end we are slaughtered with hideous cruelty'. This emphasises how badly Mankind treats animals.

❸ Pick out one of the **words** or **phrases** the student has quoted and write a further sentence to complete the explanation:

The word/phrase '..' *suggests that* ....

....

....

EXPLAINING IDEAS

You need to be precise about the way George Orwell gets ideas across. This can be done by varying your use of verbs (not just using 'says' or 'means').

❹ Read this paragraph from a **mid-level** response to a question about Napoleon's leadership. Circle all the **verbs** that are repeated in the student's writing (not in the quotation):

The book shows us how ruthless and violent Napoleon has become when he has Snowball chased away by 'nine enormous dogs'. It not only says that Napoleon will do anything to get power, and shows how he far he is prepared to go to get it, it also shows how the pigs have abandoned the principles of Major.

❺ Now choose some of the words below to replace your circled ones:

suggests | *implies* | *tells us* | *presents*
signals | *asks* | *demonstrates*
recognise | *comprehend* | *reveals* | *conveys*

❻ Rewrite your **higher-level** version of the paragraph in full below. Remember to mention the **author by name** to show you understand that he is **making choices** in how he presents characters, themes and events.

....

....

....

....

....

....

....

PROGRESS LOG [tick the correct box] Needs more work ☐ Getting there ☐ Under control ☐

Making inferences and interpretations (AO2)

WRITING ABOUT INFERENCES

You need to be able to show you can read between the lines, and make inferences, rather than just explain more explicit 'surface' meanings.

Here is an extract from one student's **very good** response to a question about Boxer and how he is presented:

At the end of Chapter Nine, just after Boxer is taken away, the narrator tells us that the pigs bought another crate of whisky with money acquired 'from somewhere or other'. By placing this incident immediately after Boxer's death, Orwell is indicating that the pigs have sold Boxer's carcass and are spending the profit on their own pleasure. It suggests that the pigs have no respect whatsoever for Boxer and his enormous contribution to the farm, and no respect for the other animals who could have benefited from the money they have spent.

1. Look at the response carefully.
   - **Underline** the simple point which explains what the pigs do.
   - **Circle** the sentence that develops the first point.
   - **Highlight** the sentence that shows an inference and begins to explore wider interpretations.

INTERPRETING – YOUR TURN!

2. Read the opening to this student response carefully and then **choose the sentence** from the list which shows **inference** and could lead to **a deeper interpretation**. Remember – interpreting is *not* guesswork!

When Clover learns that the pigs sleep in beds, she asks Muriel to read the Fourth Commandment which has now been changed. We are told that she 'had not remembered that the Fourth Commandment mentioned sheets'. This suggests that she is not intelligent enough to remember what it used to say. It also shows how …

   a) *the pigs are changing the commandments which were so important when the animals took over the farm at the beginning of the novel.*

   b) *the pigs are able to exploit the other animals' lack of intelligence and manipulate them.*

   c) *bad horses' memories are because they cannot remember anything and they struggle to understand basic ideas.*

3. Now complete this **paragraph** about Benjamin, adding your own final sentence which makes inferences or explores wider interpretations:

Benjamin is presented as cynical and bad-tempered. The only comment he makes about the rebellion is that 'Donkeys live a long time.' This suggests that …

....

....

....

PROGRESS LOG [tick the correct box] Needs more work ☐ Getting there ☐ Under control ☐

Writing about context (A03)

EXPLAINING CONTEXT

When you write about context you must make sure it is relevant to the task.

Read this comment by a student about Napoleon:

Reading 'Animal Farm' as an allegory, Napoleon can be seen as representing Stalin, the tyrannical leader of the Soviet Union. Just as Stalin forced Trotsky out of government and out of Russia, Napoleon forces Snowball to flee Animal Farm. Orwell's decision to represent Stalin as a pig may just be due to their intelligence, making a pig a more plausible leader. However, it also creates a strongly negative impression of Stalin as greedy and selfish.

1. Why is this an effective paragraph about context?
   - a) Because we learn that Stalin was a leader of the Soviet Union.
   - b) Because it reflects Orwell's political ideas and links them to character.
   - c) Because it makes us dislike Napoleon and Stalin.

EXPLAINING – YOUR TURN!

2. Now read this further paragraph, and complete it by choosing a suitable point related to context.

The farm animals' lack of intelligence is central to the success of the pigs' corrupt rule on Animal Farm. Without it, the pigs would not be able to control, mislead and manipulate them. For example ...

- a) *when Boxer tries to learn the alphabet he cannot 'get beyond the letter D'. It is not surprising then that he believes 'Napoleon is always right' and accepts everything the pigs say.*
- b) *when Squealer is found 'sprawling' beside a ladder and a pot of paint, 'None of the animals could form any idea as to what this meant.' Throughout the novel Orwell implies that the people of Russia, represented by the animals, were not only manipulated but so poorly educated that they were unable to recognise it.*
- c) *when the pigs add 'with sheets' or 'to excess' to the commandments, the animals simply accept it. None of them questions when, how or why the pigs have changed the commandment and none of them can remember what the commandment used to say. The pigs exploit the animals' poor memories to do what they like.*

3. Now, write a paragraph about how Orwell uses the final chapter to comment on the political situation in the 1940s (for example, the animals cannot tell the pigs and the men apart).

Orwell shows how ... .

....

....

....

PROGRESS LOG [tick the correct box] Needs more work ☐ Getting there ☐ Under control ☐

Structure and linking of paragraphs (A01) (A04)

Your paragraphs need to demonstrate your points clearly by:

- Using topic sentences
- Focusing on key words from quotations
- Explaining their effect or meaning

❶ Read this model paragraph in which a student explains how George Orwell presents Mollie.

Orwell presents Mollie as vain, foolish and idle. The narrator explains that every morning she 'complained of mysterious pains, although her appetite was excellent.' The use of the word 'mysterious', and the comment on her appetite, strongly suggest the narrator's opinion that she is lying in order to avoid work.

Look at the response carefully.

- **Underline** the topic sentence which explains the main point about Mollie.
- **Circle** the word that is picked out from the quotation.
- **Highlight** or put a **tick** next to the part of the last sentence which explains the word.

❷ Now read this **paragraph** by a student who is explaining how Orwell presents Clover:

We find out about Clover when the narrator describes her looking after some ducklings. 'Clover made a sort of wall around them with her great foreleg, and the ducklings nestled down inside it.' This tells us that she is kind and caring.

Expert viewpoint: This paragraph is unclear. It does not begin with a topic sentence to explain how Orwell presents Clover and doesn't zoom in on any key words that tell us what Clover is like.

Now **rewrite the paragraph**. Start with a **topic sentence**, and pick out a **key word or phrase** to **'zoom in'** on, then follow up with an explanation or interpretation.

Orwell presents Clover as … ....

....

....

....

....

....

....

....

....

....

....

....

....

....

It is equally important to make your sentences link together and your ideas follow on fluently from each other. You can do this by:

- Using a mixture of short and long sentences as appropriate
- Using words or phrases that help connect or develop ideas.

❸ Read this model paragraph by one student writing about the theme of equality:

Orwell constantly draws the reader's attention to the lack of equality on Animal Farm and so highlights how the pigs have abandoned Old Major's principles. This is perhaps clearest when the pigs re-write the most important commandment: 'All animals are equal but some are more equal than others.' Although this is obviously illogical and extremely unfair, all the animals accept it. However, this is the pigs' most honest admission. It is the first time that they have admitted any kind of inequality on the farm, suggesting that from this point on the situation will only get worse.

Look at the response carefully.

- **Underline** the topic sentence which introduces the main idea.
- **Circle** the short sentence which signals a change in ideas.
- **Highlight** or put a **tick** next to any words or phrases that link ideas, such as 'who', 'when', 'suggesting', 'which', etc.

❹ Read this **paragraph** by another student also commenting on how Mr Jones is presented:

Orwell gives us a clear picture of Mr Jones. This can be found at the very beginning of the novel. He describes him as 'too drunk' to lock up the chickens. This gives the impression that he is not a responsible person. It suggests he does not care about the animals. Orwell makes us unsympathetic to him. This makes us more sympathetic to the animals.

Expert viewpoint: The candidate has understood how the character's nature is revealed in his actions. However, the paragraph is rather awkwardly written. It needs improving by linking the sentences with suitable phrases and joining words such as: 'where', 'in', 'as well as', 'who', 'suggesting', 'implying'.

Rewrite the **paragraph**, improving the **style**, and also try to add a **concluding sentence** summing up the character of Mr Jones.

Start your **topic sentence** in the same way, but extend it:

Orwell gives us a clear picture of Mr Jones ... .

....

....

....

....

....

....

....

....

....

PROGRESS LOG [tick the correct box] Needs more work ☐ Getting there ☐ Under control ☐

Spelling, punctuation and grammar

Here are a number of key words you might use when writing in the exam:

Content and structure	Characters and style	Linguistic features
allegory	character	language
chapter	role	emotive
quotation	protagonist	juxtaposition
sequence	dramatic	dramatic irony
dialogue	satire	repetition
climax	tyrannical	symbol
development	humorous	rhetorical
introduction	sympathetic	persuasive

❶ Circle any you might find difficult to spell, and then use the 'Look, Say, Cover, Write, Check' method to learn them. This means: **look** at the word; **say** it out loud; then **cover** it up; **write** it out (without looking at it!); uncover and **check** your spelling against the correct version.

❷ Create a **mnemonic** for five of your difficult spellings. For example:

allegory: **a**ll **l**arge **l**ions **e**njoy **g**ravy **o**n **r**oast **y**ak! Or …

break the word down: AL – LEG – ORY!

a) ……

b) ……

c) ……

d) ……

e) ……

❸ Circle any **incorrect spellings** in this paragraph and then rewrite it:

In Chapter Too, the animals carry out the rebbelian and sucessfuly drive Mr and Mrs Jones from the farm. They have the best harvest ever and there work goes 'like clockwork'. Orwell encurrages the reader to enjoy their sucess, making the later dissent into tyrrany all the more shocking and upseting.

❹ **Punctuation** can help make your meaning clear. Here is one response by a student commenting on the structure of *Animal Farm*. Check for incorrect use of:

- Apostrophes, full stops, commas and capital letters
- Speech marks for quotations and emphasis

The final sentence of Orwells Animal Farm is the most disturbing. He describes the animal's looking from pig to man but they cannot say 'which was which. We are witnessing not just a strange transformation but the final nail in the coffin of Animalism, the pigs have become what they once claimed to hate.

Rewrite it **correctly** here:

....

....

....

....

....

....

❺ It is usually better to use the **present tense** to describe what is happening in the text.

Look at these two extracts. Which one uses tenses **consistently** and **accurately**?

Student A: *Napoleon claimed to have no interest in the windmill. He even urinated on Snowball's plans. However, when he drove Snowball out of Animal Farm, he announced that it was his idea all along. Suddenly it is revealed that Napoleon only wanted to have the glory of the project for himself.*

Student B: *Napoleon claims to have no interest in the windmill. He even urinates on Snowball's plans. However, when he drives Snowball out of Animal Farm, he announces that it had been his idea all along. Suddenly it is revealed that Napoleon only wants to have the glory of the project for himself.*

❻ Now look at this further paragraph. **Underline** or **circle** all the **verbs** first.

Later, Napoleon changed the facts and told the other animals that Snowball was a traitor who spied for Jones. In this way Orwell highlighted how dishonest tyrants used scapegoats to take the blame for their mistakes and will create a common enemy to fear.

Now rewrite it using the **present tense** consistently:

....

....

....

....

....

....

....

....

....

PROGRESS LOG [tick the correct box] Needs more work ☐ Getting there ☐ Under control ☐

Tackling exam tasks

A01 A02

DECODING QUESTIONS

It is important to be able to identify key words in exam tasks and then quickly generate some ideas.

1. Read this task and notice how the key words have been underlined.

Question: *How is the effect of fear and violence shown in* Animal Farm?

Write about:

- How the pigs create fear and use violence
- How Orwell presents the animals' response to fear and violence by the ways he writes.

Now underline the key words in this task:

Question: *How does George Orwell explore ideas about power in* Animal Farm*?*

Write about:

- How power is gained, used and kept in the novel
- How Orwell presents those ideas

GENERATING IDEAS

2. Now you need to generate ideas quickly in response to the question above. Use the spider diagram* below and add as many ideas of your own as you can:

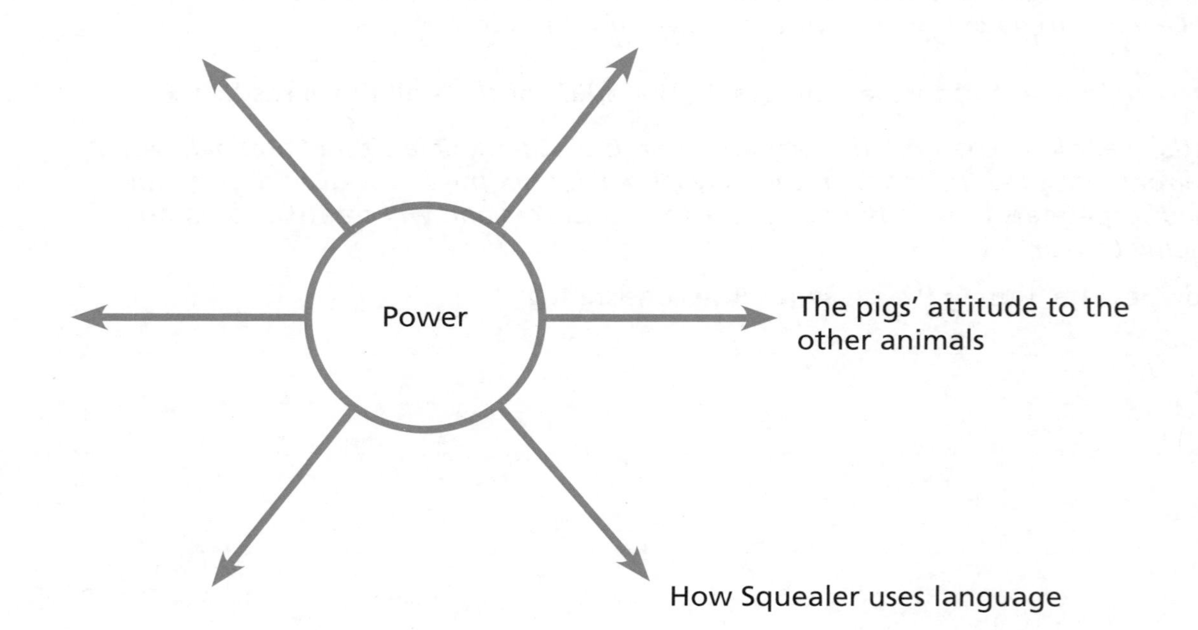

*You can do this as a list if you wish.

PLANNING AN ESSAY

❸ **Using the ideas you generated** in the spider diagram or list for Question 2, write a simple **plan** with at least **five key points** (the first two have been done for you).

a) *The pigs lie to and manipulate the other animals.*

b) *Squealer uses language to deceive the animals and create fear.*

c) ....

....

d) ....

....

e) ....

....

❹ Now list **five quotations**, one for each point (the first two have been provided for you):

a) *'"Never mind the milk, comrades," cried Napoleon, placing himself in front of the buckets.'*

b) *'Jones would come back! Yes, Jones would come back!'*

c) ....

....

d) ....

....

e) ....

....

❺ Now read this task and **write a plan of your own**, including **quotations**.

Read from *'There was a deadly silence.'* (Ch. 10, p. 84) to *'which Mrs Jones had been used to wear on Sundays.'* (Ch. 10, p. 85).

Question: *How are the pigs depicted in this extract and how do the animals respond to them?*

Start your plan here and continue it on a separate sheet of paper.

....

....

....

....

....

....

PROGRESS LOG [tick the correct box] Needs more work ☐ Getting there ☐ Under control ☐

Sample answers

AO1 AO2 AO3 AO4

OPENING PARAGRAPHS

Here is the task from the previous page:

Question: *How are the pigs depicted in this extract and how do the animals respond to them?*

Now look at these two alternative openings to the essay and read the expert viewpoints underneath:

Student A

> *Orwell depicts the pigs as sinister and ridiculous when they appear walking on two legs. Throughout the novel they have become more and more like men and this is the final, disturbing stage of that change. Perhaps more disturbing though is the way in which Orwell suggests that when the animals might be about to 'protest' at the sight of the pigs on two legs, the sheep, trained by the pigs for just this purpose, silence them with their chanting.*

Student B

> *The sight of the pigs walking on two legs is strange and when they start wearing clothes that makes it even stranger. The animals notice that the commandments have been changed again but they do not comment on it and do not question it. Benjamin reads the commandment to Clover whose sight is failing which makes her sound old. The sheep stop any protests though.*

Expert viewpoint 1: This opening makes a limited point, using and repeating informal language choices, and does not fully develop it. It soon begins to recount the story of the extract without analysis and without making it clear how this is relevant to the question.

Mid level

Expert viewpoint 2: This is a clear opening paragraph that outlines some of the ideas and their impact, which will be discussed in more detail in the main body of the essay. The depiction of the pigs could have been mentioned more fully, for example introducing ideas about how and why Orwell may intentionally be making them look ridiculous.

1. Which comment belongs to which answer? Match the paragraph (A or B) to the expert's feedback (1 or 2).

   Student A: .... **Student B:** ....

2. Now it's your turn. Write the opening paragraph to this task on a separate sheet of paper:

Read from *'It was about this time'* (Ch. 6, p. 42) to *'no complaint was made about that either.'* (Ch. 10, p. 43).

Question: *How are the pigs depicted in this extract and how do the animals respond to them?*

Remember:

- Introduce the topic in general terms, perhaps **explaining** or **'unpicking'** the key **words** or **ideas** in the task (such as 'depicted').
- Mention the **different possibilities** or ideas that you are going to address.
- Use the **author's name**.

WRITING ABOUT TECHNIQUES

Here are two paragraphs in response to a different task, where the students have focused on the writer's techniques. The task is:

Read from *'It was about this time'* (Ch. 6, p. 42) to *'no complaint was made about that either.'* (Ch. 10, p. 43).

Question: *What techniques does Orwell use to show how the pigs control the other animals?*

Student A

Orwell gives Squealer a sequence of rhetorical questions which the animals do not, or perhaps dare not, interrupt. This is because the questions need no answer: the answers are obvious. None of the animals 'wishes to see Jones back' and so Squealer exploits their fears to justify the pigs' actions and uses the rhetorical questions to make the pigs' decisions seem inevitable. Their authority cannot be challenged.

Student B

Squealer uses a series of rhetorical questions to persuade the animals that the pigs need to sleep in beds so that they can manage 'all the brainwork' that they do. These rhetorical questions end with 'Surely none of you wishes to see Jones back?' which shows how Squealer uses fear to frighten the animals into silence. This is what Squealer does very well in the novel.

Expert viewpoint 1: This higher-level response explores the immediate impact of the language techniques used by Squealer and explores how they are effective and their wider implications in the context of the novel. The penultimate sentence is a little long, but links ideas very successfully and is effectively summarised in the final short sentence.

Higher level

Expert viewpoint 2. This mid-level response highlights Squealer's use of rhetorical questions on the other animals and comments on the effect of one. However, this comment could be further developed and focus more broadly on why Orwell chose to use this technique at this point in the novel.

Mid level

3. Which comment belongs to which answer? Match the paragraph (A or B) to the examiner feedback (1 or 2).

   Student A: .... **Student B:** ....

4. Now, take another **aspect** of the passage and write your own **paragraph**. You could **comment** on one of these aspects:

- How Orwell implies the truth behind the pigs' lies
- Squealer's manipulation of the animals using false arguments
- Orwell's use of irony when describing Clover reading the Fourth Commandment

Start your plan here and continue it on a separate sheet of paper.

....

....

....

Now read this **lower-level** response to the following task:

Read from *'As the human beings approached'* (Ch. 4, p. 25) to *'had already recovered and made off.'* (Ch. 4, p. 27).

Question: *How is Man depicted in this extract?*

Student response

When the men arrive on the farm, the animals all work together to defeat them. At first the men think they have won: 'a shout of triumph.' Because we're on the animals' side we feel disappointed but that soon changes when the men are fooled and Jones ends up humiliated: 'hurled into a pile of dung.' This makes the men look silly because they are easily fooled by a bunch of animals.

After this the animals are even more successful as they all attack and defeat the men in five minutes. Maybe we feel a bit sorry for the boy who Boxer thinks he has killed but then it turns out he's OK so we feel more sorry for Boxer because he cries about it.

Expert viewpoint: The quotations in the first paragraph are well chosen and give us a sense of how the men are presented, but there is no attempt to embed them in a sentence. Nor is there any exploration in either paragraph of the effect of Orwell's choices. Further development of comments is needed, and the language the student uses is often too informal, as in, 'look silly', 'a bunch of animals'.

❺ **Rewrite** these two **paragraphs**, improving them by addressing:

- The lack of development of linking of points – no **'zooming in'** on **key words and phrases**
- The lack of **quotations and embedding**
- Unnecessary **repetition**, poor **specialist terms** and use of **vocabulary**

Paragraph 1:

In this scene, Orwell depicts the men as ....

....

and also ....

....

This implies that ....

....

....

Paragraph 2:

Our response changes when we learn that ....

....

However ....

....

This links to ....

....

....

A FULL-LENGTH RESPONSE

⑥ Write a full-length response to this exam-style task on a separate sheet of paper. Answer both parts of the question:

Question: *How is Squealer depicted by George Orwell throughout the novel?*

Write about:

- How Orwell presents him through what he does
- How Orwell presents him through what he says

Remember to do the following:

- Plan **quickly** (no more than 5 minutes) what you intend to write, jotting down **four or five supporting quotations**.
- Refer closely to the **key words** in the question.
- Comment on **what** the writer does, the **techniques** he uses and the **effect** of those techniques.
- Support your points with **well-chosen quotations** or other evidence.
- Develop your points by **'zooming in'** on particular **words** or **phrases** and explaining their **effect**.
- Be **persuasive** and **convincing** in what you say.
- Carefully check your **spelling**, **punctuation** and **grammar**.

PROGRESS LOG [tick the correct box] Needs more work ☐ Getting there ☐ Under control ☐

Further questions

A01 A02 A03 A04

① Why do you think George Orwell wrote *Animal Farm*?

② Which character is presented most negatively in the novel and why?

③ Read from *'In the long pasture'* (Ch. 5, p. 30) to *'without uttering a word.'* (Ch. 5, p. 31). How are the characters of Snowball and Napoleon, and their relationship, depicted in this extract?

④ The novel is carefully structured. What part do the Seven Commandments play in this?

⑤ How does Orwell explore the theme of education in *Animal Farm*?

PROGRESS LOG [tick the correct box] Needs more work ☐ Getting there ☐ Under control ☐

ANSWERS

Note: Answers have been provided for most tasks. Exceptions are 'Practice tasks' and tasks which ask you to write a paragraph or use your own words or judgment.

PART TWO: PLOT AND ACTION

Chapter 1 (pp. 8–11)

1 c)

2 a) F; b) T; c) T; d) T; e) F; f) T

3 a); c); d); e); g)

4 Moses

5 a) They would never be overworked or hungry; they would be working for themselves and become 'rich and free' (p. 5).

b) As Major says this, the dogs and cats chase the rats back to their holes, suggesting that the animals may struggle to overcome their natural differences and work together.

c) All the animals (except Moses) come to the meeting, listen respectfully to his speech, and sing 'Beasts of England' with him.

d) In addition to the animals' regard for Major and, consequently, his ideas, the animals show their support for Major's declaration that 'All animals are comrades' (p. 5) by voting with 'an overwhelming majority ' (p. 6) that the rats are to be considered comrades.

6 a) Boxer is enormously strong and powerful, is respected for his 'steadiness of character' (p. 2) but lacks great intelligence.

b) Benjamin is the oldest animal on the farm. He is bad-tempered, talks little, does not laugh at all, and is cynical. He is, though, 'devoted to Boxer' (p. 2).

c) Clover is 'motherly' (p. 2) as she protects the orphan ducklings from being trampled by the larger animals.

d) Mollie is 'pretty' but 'foolish' (p. 2) and seems vain as she shows off the red ribbons in her mane.

7 'Beasts of England' is similar in many ways to a national anthem: a song that sets out the aims and ambitions of the animals to focus them on their goal. Singing it in unison brings them together and reinforces their unity. It also allows Orwell to highlight the differences between the animals: the pigs and dogs learn it quickly; the 'stupidest' (p. 8) animals pick up the tune but only a few of the words.

8

Point/detail	Evidence	Effect or explanation
1: *Major describes the lives of farm animals.*	*'Let us face it: our lives are miserable, laborious and short.'*	*Major's blunt, negative language in this short sentence emphasises how badly farm animals are treated.*
2: *Major explains how mankind exploits animals.*	*'our labour is stolen from us by human beings'*	*Major implies that the animals do all the work while Man reaps all the benefits.*
3: *Major describes the deaths of farm animals.*	*'we are slaughtered with hideous cruelty'*	*Major's emotive vocabulary choices emphasise the mistreatment of animals from birth to death.*

Chapter 2 (pp. 12–15)

1 b)

2 a)

3 b); c)

4 A 5, B 8, C 1, D 7, E 9, F 2, G 6, H 3, I 4

5 a) Napoleon is 'fierce-looking' (p. 9) and likes to get 'his own way' (p. 9). He is 'not much of a talker' (p. 9) and says very little in Chapter 2, except to send the other animals to get in the harvest while he remains with the milk, which then disappears. Orwell presents him as a strong, silent but self-interested leader.

b) Snowball is much more 'vivacious' (p. 9) than Napoleon and quick to respond to the other animals' concerns. Much of the leadership given by the pigs in the chapter is voiced by Snowball.

c) Squealer is a brilliant, persuasive speaker who can turn 'black into white' (p. 9).

6 Sugarcandy Mountain suggests a form of 'heaven' and offers a kind of religious support to the animals, promising a paradise after death to compensate for a miserable life on Earth. Moses shares his name with the Christian prophet from the Old Testament of the Bible, who brought the Ten Commandments from God to men.

7 While the reader may ultimately blame Jones for the negligence of his farm, Orwell does give some reasons to excuse him – lack of money and loss of heart, suggesting that he does not want the reader to hold him entirely responsible.

8 The animals' objections suggest a lack of intelligence, vision, optimism and determination.

9 The word 'disciples' has connotations of religion and religious followers, suggesting the horses regard Animalism as a religion, and the pigs as the farm's saviours.

10 Major clearly told the animals that they should not behave like Man and that 'we must not come to resemble him' (p. 6).

11 The implication that Napoleon steals the milk and deprives the other animals of it suggests that he has already broken the 7th Commandment: 'All animals are equal' (p. 15). It could also be argued that Mollie trying on a ribbon in Jones's bedroom (p. 14) is breaking the 3rd Commandment: 'No animal shall wear clothes.' (p. 15).

12

Point/detail	Evidence	Effect or explanation
1: *The pigs use their intelligence to control and direct the other animals.*	*'The pigs now revealed that during the past three months they had taught themselves to read and write'*	*The word 'revealed' indicates that the pigs have acquired this human skill secretly and may not be honest or fair in their running of the farm.*
2: *Orwell clearly implies why the milk has disappeared at the end of the chapter.*	*'"Never mind the milk, comrades," cried Napoleon placing himself in front of the buckets.'*	*From the very start of the rebellion, Napoleon's selfishness and corruption is made clear.*
3: *Orwell suggests that the pigs may not look after the animals as well as Jones did.*	*'Jones used sometimes to mix some of it [the milk] in our mash'*	*While Jones shared the milk with the hens, Napoleon does not, highlighting his self-interest. The rebellion already seems unlikely to bring any improvement in the lives of the animals.*

Chapter 3 (pp. 16–17)

1 a) The pigs adapt the tools; they do not work but 'directed and supervised' (p. 16).

b) The horses pull the machinery.

c) The hens and ducks gather up stalks.

d) All the other animals turn and gather the hay.

2 b)

3 a) Benjamin may be suggesting that, with his long life and experience, he can see more clearly how good intentions can be corrupted or exploited and that the farm will be no better under the pigs than it was under Jones.

b) The pigs are becoming more corrupt and putting their own interests above those of the other animals. Already, all animals are *not* equal.

c) Squealer uses the false premise that the pigs' intelligence relies on eating apples and drinking milk. It is an effective but dishonest argument as he perhaps knows that the less intelligent animals will not question it and that it plays upon their greatest fear.

4

Point/detail	Evidence	Effect or explanation
1: *The pigs' intelligence allows them to manipulate and exploit the farm animals just as Jones did.*	*'The birds did not understand Snowball's long words, but they accepted his explanation'*	*The pigs are effectively able to control the other animals' thoughts, attitudes and decisions.*
2: *The pigs take all the apples and milk for themselves.*	*'The animals had assumed … that these would be shared out equally'*	*While the animals expect to live by the principles of Animalism, the pigs ignore those principles.*
3: *Squealer's dishonest explanation suggests the pigs will continue to lie and cheat to get their own way.*	*'Jones would come back!'*	*Squealer plays upon the animals' worst fear – although the pigs are treating them more and more as Jones did.*

Chapter 4 (pp. 18–19)

1 b); c)

2 b)

3 a)

4 a) Snowball and Napoleon's motives could be positive: to free all animals from human control; however, they could be attempting to bring more animals under their control so that they can exploit them as they are already doing on their own farm.

b) Orwell may have wanted to maintain the reader's sympathy for the animals. The murder of a human could have jeopardised this.

c) The tone of the battle – in which the pigeons 'muted' (p. 25) on the men and Jones is 'hurled into a pile of dung' (p. 26) – and the animals' 'wildest excitement' (p. 27) at their victory suggests that the reader should enjoy the battle and the animals' triumph.

5

Point/detail	Evidence	Effect or explanation
1: *Snowball is wounded.*	*'from whose wounds the blood was still dripping'*	*The fact that Snowball is wounded suggests his dedication to Animalism and the other animals.*
2: *The animals work together very successfully.*	*'each recounting his own exploits'*	*Despite their differences, the animals are able to work together in a crisis.*
3: *The animals celebrate their victory.*	*'wildest excitement'*	*The animals' cooperation and success could signal a more positive mood as all the pigs except Napoleon fight alongside the other animals.*

Chapter 5 (pp. 20–1)

1 a)

2 Napoleon: The animals should concentrate on food production, not windmills; The animals should get guns to protect themselves from another attack by humans.

Snowball: The windmill would save so much labour, the animals could work for just three days a week; The animals should stir up rebellion on other farms to make all animals more powerful.

3 a) The sheep's bleating interrupts Snowball's speech but not Napoleon's. Later events suggest that Napoleon is using the sheep in an attempt to weaken or drown out Snowball.

b) Later events, again, suggest that Napoleon has no need to give his ideas or argue against Snowball's; he knows he will soon use brute force and violence to silence Snowball and intimidate the other animals.

c) This suggests that Napoleon only criticised the windmill because it was Snowball's idea and worried that it would increase Snowball's influence and power. Napoleon is now able and happy to take the credit for it.

4

Point/detail	Evidence	Effect or explanation
1: *Orwell presents the dogs as vicious and dangerous.*	*'a terrible baying sound … enormous dogs … snapping jaws'*	*Focusing on their size and their jaws, Orwell highlights how intimidating the dogs are, and by implication how intimidating Napoleon is too.*
2: *The animals are deeply affected by this incident.*	*'Silent and terrified'*	*Orwell suggests that this was Napoleon's intention: to frighten the animals into silence so his leadership is not questioned.*
3: *Napoleon uses this incident to increase his power.*	*'there would be no more debates.'*	*Orwell emphasises that Napoleon's dictatorship means an end to democracy.*

Chapter 6 (pp. 22–3)

1 c)

2 a); c); d)

3 a) Orwell presents Boxer as the most selfless and tireless of all the animals – which makes events later in the text all the more shocking and disturbing.

b) Squealer's main role is to support Napoleon's flagrant disregard for the principles and Commandments of Animalism by misleading the other animals with his eloquence and dubious arguments.

c) Snowball becomes the scapegoat of Animal Farm: blaming him for everything that goes wrong provides the animals with a common enemy to unite them and diverts attention and any possible blame away from Napoleon.

4

Point/detail	Evidence	Effect or explanation
1: For the first time, the animals are beginning to question the pigs' actions.	*'Clover … thought she remembered a definite ruling against beds'*	*Because Clover only 'thinks' she remembers, the pigs are able to exploit her poor memory.*
2: The pigs alter the commandments for their own ends.	*'No animal shall sleep in a bed with sheets'*	*A simple and irrelevant change allows the pigs to flout the commandments.*
3: Squealer verbally manipulates the other animals.	*'We … sleep between blankets … Surely none of you wishes to see Jones back?'*	*Squealer uses ridiculous and irrelevant arguments which the animals are not capable of arguing against, and which frighten them into submission.*

Chapter 7 (pp. 24–5)

1 a); b)

2 a); b); c)

3 Any three of: he steals corn, upsets milk-pails, breaks eggs, tramples seed-beds, gnaws bark from fruit trees, breaks windows, blocks drains, steals keys, milks the cows, offers to act as Frederick's guide during an attack, spies for Jones, tries to lose the Battle of the Cowshed, encourages other animals to work against Animal Farm.

4 a) Snowball's tactical retreat is now presented as surrender; Napoleon, who was absent from the battle, is now presented as taking over the leadership after Snowball's surrender and biting Jones.

b) The singing of 'Beasts of England' effectively expresses the disappointment of the animals' hopes when they first sang it, before and immediately after the rebellion, and the horrific reality of the farm on which they now live. It effectively summarises Clover's thoughts ('though she lacked the words to express them' p. 55) in the paragraph preceding its singing.

c) 'Beasts of England' reminds the animals of Major, his principles and the vision he offered the animals in Chapter 1. Singing it therefore undermines Napoleon's leadership and the animals' faith in him.

5

Point/detail	Evidence	Effect or explanation
1: Napoleon presents himself as heroic.	*'wearing both his medals … with his nine huge dogs frisking round him.'*	*Napoleon wears the medals he has awarded himself to win the animals' respect and uses his guard dogs to encourage their fear.*
2: The number of deaths and their description is horrific.	*'tore their throats out'*	*This graphic description, leaving no doubt about Napoleon's ruthlessness, is shocking and unexpected for the reader and the animals.*
3: Orwell compares the rule of Napoleon and the rule of Jones.	*'the smell of blood, which had been unknown there since the expulsion of Jones.'*	*Orwell is clearly suggesting that this is no different from, and no improvement upon, when Jones ran the farm.*

Chapter 8 (pp. 26–7)

1 All of them.

2 a) F; b) NEE; c) NEE; d) NEE; e) F; f) F

3 a) He has a cockerel as his personal 'trumpeter'; he eats alone, waited on by dogs; he has his own apartment in the farmhouse; a gun will be fired on his birthday; he is referred to as 'Our Leader, Comrade Napoleon'; the pigs invent titles for him e.g. 'Father of All Animals'; Minimus writes a poem in his honour; his portrait is painted on the barn; the windmill is named after him; he invents and awards himself the 'Order of the Green Banner'.

b) The hat suggests that Napoleon is becoming more and more like Jones – and has broken one of the Commandments in doing so.

4

Point/detail	Evidence	Effect or explanation
1: The pigs drink Jones's alcohol in the farmhouse.	*'Napoleon was distinctly seen to … gallop rapidly round the yard'*	*The pigs' privilege and disregard for the commandments is growing.*
2: The pigs plough up the retirement pasture to grow barley for brewing alcohol.	*'It was given out that the pasture was exhausted'*	*The pigs are becoming increasingly and flagrantly dishonest and heartless.*
3: Orwell strongly implies that Squealer has altered the Commandments.	*Squealer is found 'sprawling' near 'a paint-brush and an overturned pot of white paint'*	*Orwell is suggesting more and more overtly that the pigs are manipulating the other animals.*

Chapter 9 (pp. 28–9)

1 c)

2 b); c)

3 a) It suggests the superiority of the pigs. In Orwell's lifetime, servants were expected to do the same when they met their masters; some were even expected to stand perfectly still, facing the wall!

b) The money will benefit the pigs as most of it will be spent on supplies for the farmhouse. The rest of the animals will suffer as the money is made by, effectively, reducing their food rations.

c) It is hard to see the animals under Napoleon's regime as anything other than slaves. They are free in name only; they have no freedom to think, form opinions, make decisions, etc.

4

Point/detail	Evidence	Effect or explanation
1: Boxer has no idea of the fate that awaits him.	*'he looked forward to the peaceful days … in the corner of the big pasture.'*	*The reader already suspects that retirement is another of the pigs' lies and so tension builds as we wait to see what happens to Boxer.*
2: The animals have no idea where Boxer is going.	*'"Good-bye, Boxer!" they chorused, "good-bye!"'*	*Again, the animals' naivety and lack of intelligence encourages the reader's sympathy.*
3: Only Benjamin sees what is happening.	*'Alfred Simmonds, Horse Slaughter and Glue-Boiler'*	*By revealing the truth in this graphic language, Orwell maximises the readers' shock and disgust.*

ANSWERS

Chapter 10 (pp. 30–1)

1 a); b); c)

2 a); c)

3 a) The pigs' role is to do paperwork: to manage the farm, not do any of the actual farm work.

b) The most likely responses are amusement at such a ridiculous sight, and shock that the pigs have so completely tried to become like Man.

c) It suggests that there is now no difference at all between pigs and men: the farm is being run, not on the principles of Major, but on the principles of Man.

4

Point/detail	Evidence	Effect or explanation
1: *Little has changed in the lives of the animals.*	*'They were generally hungry … they laboured in the fields.'*	*It is now several years later; nothing has changed and it now seems unlikely ever to change.*
2: *The animals still believe in the principles of Animalism.*	*'the animals never gave up hope'*	*Their hope seems even more naive and futile than ever.*
3: *The animals believe everything they are told, even though they have no evidence to support it.*	*'All animals were equal.'*	*Orwell emphasises the great irony of this mistaken belief by placing it in its own short sentence.*

PART THREE: CHARACTERS

Who's who? (p. 33)

From left to right: Mr Jones; Major; Mr Pilkington & Mr Frederick; Mr Whymper; Napoleon; Squealer; Boxer; Moses; Benjamin; Clover; Muriel; Snowball; Mollie

Major (p. 34)

1/2 wise (Ch. 1, p. 1); respected (Ch. 1, p. 1); intelligent (Ch. 1, p. 1); visionary (Ch. 1, p. 5); influential (p. 9 and throughout Ch. 2)

3 Major's ideas are adopted by the animals and organised by the pigs. They are then reduced and corrupted by the pigs in order that they can control and manipulate the animals.

4 a) Major inspired Animalism through his speech; b) Major argues that Man exploits animals; c) Major urges the animals to rebel.

Snowball (p. 35)

1 a) T; b) F; c) F; d) T

2 a) 'Snowball was a more **vivacious** pig than Napoleon, quicker in **speech** and more **inventive**.' (Ch. 2, p. 9)

b) 'Then Snowball (for it was Snowball who was best at **writing**) took a brush between the two knuckles of his trotter, painted out **Manor Farm** from the top bar of the gate and in its place painted **Animal Farm**.' (Ch. 2, p. 14)

c) 'The birds did not understand Snowball's **long words**, but they accepted his **explanation**' (Ch. 3, p. 21)

d) 'The only **good** human being is a **dead** one.' (Ch. 4, p. 26)

e) 'One of them all but closed his **jaws** on Snowball's **tail**, but Snowball whisked it free just in time.' (Ch. 5, p. 33)

Napoleon (p. 36)

Quality	Moment/s in story	Quotation
a) Arrogant	*He has a cockerel who walks ahead of him announcing his arrival.*	*'a black cockerel who marched in front of him and acted as a kind of trumpeter' (Ch. 8, p.57)*
b) Ruthless	*He executes numerous animals.*	*'They were all slain on the spot.' (Ch. 7, p. 53)*
c) Selfish	*He drinks all the milk.*	*'it was noticed that the milk had disappeared.' (Ch. 2, p. 16)*
e) Tyrannical	*He has Snowball chased away.*	*'They dashed straight for Snowball, who only sprang from his place just in time to escape their snapping jaws.' (Ch. 5, p. 33)*

2 dictatorial, hypocritical, lazy, manipulative, devious, cowardly, intelligent, shrewd, disloyal

3 a) Orwell infers that Napoleon's chief motivations are self-interest and a desire for power. He ruthlessly maintains his power by whatever means are necessary and has no qualms about breaking and re-writing the commandments in order to make his life as comfortable as possible.

b) Napoleon's use of extreme and heartless violence, e.g. the executions, or the death of Boxer, are shocking, however, it could be argued that his disregard for the principles of Animalism and transformation into Man is yet more disturbing.

Squealer (p. 37)

1 a) T; b) T; c) NEE; d) T; e) F

2

Tactic	Example
Lies	*'Many of us actually dislike milk and apples.' (Ch. 3, p. 22)*
Fear	*'Jones would come back!' (Ch. 3, p. 22)*
False facts and statistics	*'this has been proved by Science, comrades' (Ch. 3, p. 22)*
False arguments	*'It is for your sake that we drink that milk and eat those apples. Do you know what would happen if we pigs failed in our duty?' (Ch. 3, p. 22)*
Rhetorical questions	*'Surely there is no one among you who wants to see Jones come back?' (Ch. 3, p. 22)*
Repetition	*'Jones would come back!' (Ch. 3, p. 22)*

Boxer (p. 38)

1 a) Boxer is enormously strong but not very intelligent. For example *he works incredibly hard to build the windmill but cannot learn more than four letters of the alphabet.*

b) Orwell emphasises how hard Boxer works when *he gets up earlier and works later and later as time goes on.*

c) Boxer's bravery is shown when he *fights in the Battle of the Cowshed.*

d) Boxer's honesty and sensitivity are shown in the Battle of the Cowshed when he *feels regret and guilt because he thinks he has killed one of the men.*

2

Quotation	Explanation
Major warns him, 'the very day that those great muscles of yours lose their power, Jones will send you to the knacker'	*This prediction – which comes true under Napoleon's rule – suggests how little has changed by the end of the text.*
Boxer's answer to everything is 'I will work harder.'	*Boxer's hard work and lack of intelligence is exploited by his rulers, who dispose of him as soon as he can no longer work.*
Boxer believes that 'Napoleon is always right.'	*Boxer represents those who accept politicians' ideas without question – and face the unpleasant consequences.*
After he is injured, 'He looked forward to the peaceful days that he would spend in the corner of the big pasture.'	*Boxer's naivety makes him an extremely sympathetic character, so accentuating the reader's sympathy for him – and antipathy towards the pigs who have him slaughtered*

Clover (p. 39)

1 a) The reader knows that Clover is slightly more intelligent than Boxer because she *learns the whole alphabet but cannot put words together. Boxer is only able to learn four letters of the alphabet.*

b) After the executions, Clover cannot express her thoughts and feelings so she *sings 'Beasts of England'.*

c) When Boxer is hurt, she *stays with him and cares for him.*

d) When Clover sees Squealer walking on two legs, she *neighs in terror.*

2

Quality	Moment/s in story	Quotation
a) Kind and caring	*She protects the ducklings.*	*'Clover made a sort of wall around them with her great foreleg' (Ch. 1, p. 2)*
b) Unintelligent	*She blames her poor memory rather than the pigs for the changes she notices in the commandments.*	*'Clover had not remembered that the Fourth Commandment mentioned sheets' (Ch. 6, p. 42)*
c) Questioning	*She questions what has happened to Major's vision.*	*'this was not what they had aimed at when they had set themselves years ago to work for the overthrow of the human race.' (Ch. 7, p. 54)*
d) Loyal	*Despite questioning what has happened to Major's vision, she does not blame the pigs.*	*'There was no thought of rebellion or disobedience in her mind.' (Ch. 7, p. 55)*

Benjamin (p. 40)

1 1C; 2A; 3B

2 A Despite his cynical nature, Benjamin is very kind to Boxer as he keeps him company when he is unwell and is, for once, moved to action when he realises that Boxer is being taken away.

B Benjamin is clearly intelligent and perceptive but will not act upon what he sees and understands. He simply stands by and watches as Frederick's men destroy the windmill.

C Almost every time he speaks, Benjamin seems to suggest that change and the attempt to change are futile.

3 Benjamin seems to suggest in his repeated statement of 'Donkeys live a long time' that he has had much experience of change in his long life but that he has never seen any improvement in it.

Mr Jones (p. 41)

1

Point	Evidence	How Orwell wants the reader to respond
1: *Mr Jones does not look after his animals or his farm very well and is driven from his farm by his own animals.*	*Orwell tells us that Jones neglects the animals because he is 'too drunk' (Ch. 1, p. 1) and his men are 'idle and dishonest' (Ch. 1, p. 11).*	*Orwell encourages the reader to respond negatively to Jones and support the animals.*
2: *Mr Jones and the neighbouring farmers try to recapture the farm but are defeated.*	*Jones is flung into a 'heap of dung' (Ch. 4, p. 26). The animals have an 'impromptu celebration' (Ch. 4, p. 27).*	*Orwell encourages the reader to enjoy and celebrate the animals' victory at the Battle of the Cowshed.*
3: *Mr Jones sometimes used to mix milk in with the hens' mash.*	*The pigs are now taking the milk for themselves: 'the milk had disappeared' (Ch. 2, p. 16).*	*While the reader may still respond to Jones largely negatively, Orwell is now focusing our antipathy on Napoleon and the pigs.*
4: *Orwell tells us what has happened to Mr Jones.*	*'he had died in an inebriates' home' (Ch. 10, p. 80)*	*Orwell is perhaps encouraging sympathy for Jones, suggesting how devastated he was by the loss of the farm which is now run with as much cruelty, if not more, by the pigs.*

2 Orwell seems to suggest that the Tsar's rule of Russia was **careless** and **neglectful**: that he took no interest in the **well-being** of his people and no interest in the **actions** of those people whose job it was to look after them.

Moses (p. 42)

1 a) 'The pigs had an even harder struggle to counteract the **lies** put about by Moses.' (Ch. 2, p. 10)

b) 'Moses, who was Mr Jones's **especial pet**, was a **spy** and a **tale-bearer**' (Ch. 2, p. 10)

c) 'The animals hated Moses because he told **tales** and did not **work**.' (Ch. 2, p. 10)

2 a) The animals' lives are hard and laborious and so, it is implied, the thought of going to such a wonderful place after death was comforting and perhaps made them feel more able to bear their hardship.

b) The existence of such a place seems unlikely – and they do not trust Moses.

c) It is implied that the pigs feel Moses's lies encourage the animals to accept their hardship and not complain or rebel as they rebelled against Jones: 'Their lives now, they reasoned, were hard and laborious; was it not right and just that a better world should exist somewhere else?' (Ch. 9, p. 73)

Minor characters (pp. 43–4)

1 Mollie likes sugar and ribbons, dislikes hard work and does not appear to care about Animalism.

2 Orwell seems to be suggesting that the Russian middle class were **selfish** and **vain** and **uninterested** in the lives of those worse off than themselves.

3 The chief role of these dogs is to produce the puppies which Napoleon trains from birth to become his 'army'. The tendency of dogs to hunt and kill, which Napoleon makes use of, is suggested when they chase the rats.

4

Quotation	Explanation
'there was a terrible baying sound outside, and nine enormous dogs wearing brass-studded collars came bounding into the barn. They dashed straight for Snowball' (Ch. 5, p. 33)	*The dogs are highly trained and highly dangerous.*
'Squealer spoke so persuasively, and the three dogs who happened to be with him growled so threateningly, that they accepted his explanation without further questions' (Ch. 5, p. 37)	*The dogs are used to intimidate the animals and enforce the pigs' rule.*
'Napoleon himself was not seen in public as often as once a fortnight. When he did appear, he was attended ... by his retinue of dogs' (Ch. 8, p. 57)	*The dogs are used not only to intimidate but as a visual representation of power.*

5 Orwell has chosen the sheep to represent those who chant slogans unthinkingly and so, without any thought or decision, effectively support those in power. Sheep are, stereotypically, unintelligent and likely to follow each other in an unthinking flock.

6

Evidence	Adjectives
In the Battle of the Cowshed, she 'rushed forward and prodded and butted the men from every side.' (Ch. 4, p. 25)	*Brave and loyal*
'Muriel ... could read somewhat better than the dogs, and sometimes used to read to the others in the evenings from scraps of newspaper which she found on the rubbish heap' (Ch. 3, p. 20)	*Intelligent*
Muriel reads the Sixth Commandment to Clover when Clover thinks it may have been changed but Muriel does not comment on the change.	*Helpful and kind but passive*

7 a) Mr Pilkington; b) Mr Frederick; c) Mr Whymper

8 All could be applied to Pilkington and Frederick; Mr Whymper is presented as devious and greedy.

PART FOUR: THEMES, CONTEXTS AND SETTINGS

Themes (pp. 46–9)

1 a) Equality: 4 Having the same rights and opportunities as others

b) Propaganda: 1 Biased or misleading information used to support a point of view

c) Corruption: 2 Dishonest activity by those in power

d) Tyranny: 3 The cruel misuse of power

2 Likely answers: power, equality, language, education, corruption, propaganda

3 Possible answers:

a) power: Snowball is chased off the farm; the executions take place

b) equality: the pigs take all the milk and apples; the pigs move into the farmhouse

c) language/propaganda: Squealer changes the commandments; Squealer threatens the animals with Jones's return

4 a) power: Orwell seems to be suggesting that power corrupts and that corruption destroys the responsibilities of the powerful.

b) equality: Orwell seems to be suggesting that true equality for all is difficult or impossible to achieve.

c) language: Orwell seems to be suggesting that language can be used to manipulate and control others.

5 a) Major: equality. For example, Major outlines his vision of an ideal society which the animals aspire to.

b) Snowball: power and corruption. For example, despite his good intentions to improve the farm, Snowball does not challenge the pigs' corruption.

c) Napoleon: power, corruption, tyranny. For example, Napoleon epitomises the appetite for power at all costs.

d) Squealer: language. For example, Squealer is the 'voice' of the pigs, using language to control, manipulate and deceive those not intelligent or educated enough to question it.

e) Boxer: equality and education. For example, Boxer's lack of intelligence and unquestioning dedication prevent him from seeing the inequality on Animal Farm.

f) Mr Jones: power and corruption. For example, Jones is an incompetent and irresponsible 'ruler' whose expulsion from the farm seems well deserved – until he is replaced with a yet more irresponsible and intentionally cruel ruler.

6 Theme: propaganda

Possible additions:

'pleadingly': suggests Squealer playing on the animals' emotions, seeming almost to beg them to agree with him

Language used persuasively: 'Surely' suggests that Squealer cannot believe anyone could want this; emphasised through repetition.

7

Point/detail	Evidence	Effect or explanation
1: *The pigs write the Seven Commandments on the barn wall but then go on to change them to suit their own corrupt ends.*	*'All animals are equal but some animals are more equal than others.'*	*The pigs use their intelligence to control the less intelligent animals, turning the most important commandment into nonsense.*
2: *The pigs use fear and violence to control the animals.*	*'the air was heavy with the smell of blood, which had been unknown there since the expulsion of Jones.'*	*The pigs corrupt the truth by presenting the murder of innocent animals as the execution of traitors.*
3: *Squealer uses the power of language to exploit the animals' fears and manipulate the truth.*	*'Surely none of you wishes to see Jones back?'*	*Squealer repeatedly uses the unlikely threat of Jones to justify the pigs' selfish actions.*

Contexts (pp. 50–3)

1 c)

2 a)

3 c)

4 b)

5 b); c); e)

6 Great Britain/The United States of America: democratic

The Soviet Union: Communist

Italy/ Germany: fascist

7 a) Orwell seems to be suggesting that, despite any political and national differences, all political leaders look and sound the same, and have similar motivations of self-interest and a desire for power and control.

b) The totalitarian rule of Napoleon has similarities not only with that of Stalin in communist Russia but Mussolini in fascist Italy and Hitler in Nazi Germany: in particular the use of violence, fear, intimidation and propaganda to control their people and maintain their power.

c) Orwell seems to make a distinction between political ideals, as set out by Major, and the reality of politics in which self-interest outweighs ideals, as exemplified in Napoleon.

8 1C; 2I; 3D; 4A;, 5E; 6H; 7G; 8F; 9B

9

Point/detail	Evidence	Effect or explanation
1: *The ideals of the rebellion are soon lost.*	*'when they came back in the evening it was noticed that the milk had disappeared.'*	*Orwell is suggesting that the Marxist ideals of the Russian Revolution were destroyed by its leaders.*
2: *The pigs treat the other animals violently and ruthlessly.*	*'They are taking Boxer to the knacker's!'*	*Orwell uses the death of Boxer, the most loyal and sympathetic character on the farm, to reflect the ruthless cruelty of Stalin.*
3: *By the end, the animals cannot tell the pigs and men apart.*	*'already it was impossible to say which was which.'*	*This symbolically implies that there is no real difference between the leaders of the Russian Revolution and the Tsar that they overthrew.*

Settings (pp. 54–5)

1 a) Pinchfield Farm

b) The farmhouse

c) The yard

d) The barn

e) The farmhouse

f) The harness room

g) The barn

h) The quarry

i) The barn

j) Above the harness room

k) Foxwood Farm

2 a) When the pigs move into Jones's farmhouse (and sleep in his beds), it becomes increasingly apparent that there is little difference between the rule of pigs and the rule of humans.

b) The setting of the farm is, occasionally, presented as idyllic but more often as a harsh and difficult place to live. In this way, Orwell compares the possibilities which Major's dream suggested and the reality of the lives which the pigs oversee.

PART FIVE: FORM, STRUCTURE AND LANGUAGE

Form (p. 57)

1 a); c); d); e); g)

2 c)

3 a) While *Animal Farm* features elements of fairy story, the subtitle is also in keeping with the ironic tone of the novel: the contrast of what we might expect of a fairy story with the brutal satire which confronts the reader helps to exaggerate the shock of the satire.

b) Orwell implies that, like the stereotypical image of a pig, politicians are greedy and lazy.

Structure (pp. 58–9)

1 a) Things go well. *Old Major's vision is appealing, the animals succeed in their rebellion and run the farm successfully.*

b) Things start to go wrong. *The pigs begin to take advantage of their superior intelligence and corruption sets in.*

c) Things go very badly wrong. *The pigs have absolute power on Animal Farm and abuse it, much as Jones did.*

2 All of them. The reader is initially encouraged to share Major's vision and enjoy Jones' expulsion and the animals' successes. From there on, the remainder of these elements charts how far the pigs are straying from Major's vision until things have come full circle and we can no longer see the difference between the pigs and Jones.

3 a) By the end of the novel, the animals' lives are no different to how they were at the beginning. Indeed, it is strongly suggested that they are worse.

b) It could be argued that the differences between the animals is apparent from the very beginning: the pigs position themselves in front of all the other animals in the barn while Clover kindly protects ducklings from the clumsier, less thoughtful animals. However, the first clear sign of the pigs' selfishness is when the milk goes missing at the end of Chapter Two.

4

Point/detail	Evidence	Effect or explanation
1: *Each of the Seven Commandments is broken and altered.*	*'No animal shall sleep in a bed with sheets'*	*Each change to the commandments signals the pigs' growing corruption.*
2: *At the end, the pigs walk on two legs and wear human clothes.*	*'his favourite sow appeared in the watered silk dress which Mrs Jones had been used to wear on Sundays.'*	*A ridiculous but shocking image emphasising how the pigs have lost sight of Major's vision.*
3: *Repeatedly the animals do not question or comment on the pigs' actions.*	*'Napoleon is always right.'*	*Orwell builds the reader's frustration as the pigs' corruption goes unchallenged.*

ANSWERS

Language (pp. 60–2)

1 a) Contrast: Emphasising the difference between two things

b) Rhetorical question: A question that needs no answer

c) Pattern of three: A list of three items, adjectives, ideas, etc.

d) Emotive language: Language that appeals to the reader or listener's emotions

e) Repetition: Using the same word, phrase or sentence more than once

f) Statistics: Numerical facts used to prove a point

2 Irony: Saying one thing but meaning exactly the opposite

Dramatic irony: When the writer provides enough information for the reader to infer or understand more than the characters in a text

3 Contrast: 'miserable, laborious ... happiness or leisure' (Ch. 1, p. 3)

Rhetorical question: 'what is the nature of this life of ours?' (Ch. 1, p. 3)

Pattern of three: 'miserable, laborious, and short.' (Ch. 1, p. 3)

Emotive language: 'we are slaughtered with hideous cruelty.' (Ch. 1, p. 3)

Repetition: 'No animal in England ... No animal in England' (Ch. 1, p. 3)

Statistics: 'This single farm of ours would support a dozen horses, twenty cows, hundreds of sheep' (Ch. 1, p. 3–4)

4 Rhetorical question: 'surely there is no one among you who wants to see Jones come back?'

Emotive language: 'what would happen if we pigs failed in our duty?'

Repetition: 'Jones would come back! Yes, Jones would come back!'

5 All are arguable. The pigs use all three: language to deceive, violence to intimidate, and fear to control.

7 a) Neither. Although this could be considered dramatic irony once we know Boxer's fate, at this point in the novel we are not aware of it.

b) Irony

c) Dramatic irony

d) Dramatic irony

e) Irony

8 a)/b) A short sentence/simple vocabulary: 'Boxer was never seen again.'

Little or no description: 'the van was through it and rapidly disappearing'

c) Short emphatic sentences bluntly announce Boxer's exhaustion and disappearance. The absence of any emotive description or response from the other animals creates an entirely unsentimental, factual tone, in sharp contrast to the sympathy Orwell has built for Boxer throughout the novel, exaggerating the reader's shock still further.

PART SIX: PROGRESS BOOSTER

Writing skills (pp. 64–5)

2 Student A: Level – Mid

Why? *The student makes a clear point that is supported with evidence, and some explanation is given.*

Student B: Level – Higher

Why? *The student uses a focused quotation and comments on the specific and the broader effect. A developed explanation is also given.*

4/5/6 Orwell **demonstrates** how ruthless and violent Napoleon has become when he has Snowball chased away by 'nine enormous dogs'. It not only **reveals** that Napoleon will do anything to get power, and **signals** how he far he is prepared to go to get it, it also **suggests** how the pigs have abandoned the principles of Major.

Making inferences and interpretations (p. 66)

1 Simple point: first sentence; develops: second sentence; inference: third sentence

2 b)

Writing about context (p. 67)

1 a)

2 b)

Structure and linking of paragraphs (pp. 68–9)

1 Topic sentence: *Orwell presents Mollie as vain, foolish and idle.*

Quotation word: *'mysterious'*

Explains: *strongly suggest the narrator's opinion that she is lying in order to avoid work*

3 Topic sentence: *Orwell constantly draws the reader's attention to the lack of equality on Animal Farm*

Change: *However, this is the pigs' most honest admission.*

Links: and, when, Although, However, that, suggesting

Spelling, punctuation and grammar (pp. 70–1)

3 In Chapter **Two**, the animals carry out the **rebellion** and **successfully** drive Mr and Mrs Jones from the farm. They have the best harvest ever and **their** work goes 'like clockwork'. Orwell **encourages** the reader to enjoy their **success**, making the later **descent** into **tyranny** all the more shocking and **upsetting**.

4 *The final sentence of Orwell's 'Animal Farm' is the most disturbing. He describes the animals looking from pig to man but they cannot say 'which was which'. We are witnessing not just a strange transformation, but the final nail in the coffin of Animalism. The pigs have become what they once claimed to hate.*

5 Student B

6 a) Later, Napoleon <u>changed</u> the facts and <u>told</u> the other animals that Snowball <u>was</u> a traitor who <u>spied</u> for Jones. In this way Orwell <u>highlighted</u> how dishonest tyrants <u>used</u> scapegoats <u>to take</u> the blame for their mistakes and <u>will create</u> a common enemy <u>to fear</u>.

b) Later, Napoleon <u>changes</u> the facts and <u>tells</u> the other animals that Snowball <u>was</u> a traitor who <u>spied</u> for Jones. In this way Orwell <u>highlights</u> how dishonest tyrants <u>use</u> scapegoats <u>to take</u> the blame for their mistakes and <u>create</u> a common enemy <u>to fear</u>.

Tackling exam tasks (pp. 72–3)

1 <u>How</u> does <u>George Orwell</u> <u>explore</u> ideas about <u>power</u> in <u>*Animal Farm*</u>?

- <u>How</u> is <u>power</u> <u>gained</u>, <u>used</u> and <u>kept</u> in the novel
- <u>How</u> Orwell <u>presents</u> those <u>ideas</u>

Sample answers (pp. 74–7)

1 Student A: Expert viewpoint 2; Student B: Expert viewpoint 1

3 Student A: Expert viewpoint 1; Student B: Expert viewpoint 2